EXISTENTIAL PHYSICS

ALSO BY SABINE HOSSENFELDER

Lost in Math:
How Beauty Leads Physics Astray

Experimental Search for Quantum Gravity
(editor)

EXISTENTIAL
PHYSICS

A Scientist's Guide to Life's
Biggest Questions

∘ ∘ ∘

SABINE
HOSSENFELDER

VIKING

VIKING
An imprint of Penguin Random House LLC
penguinrandomhouse.com

Grateful acknowledgment is made for permission to reprint an excerpt
from "Galaxy formation efficiency and the multiverse explanation of the
cosmological constant with EAGLE simulations" by Luke A. Barnes et
al, in *Monthly Notices of the Royal Astronomical Society*, Volume 477,
Issue 3, July 2018, Pages 3727–3743. Used by permission of Oxford
University Press and Dr. Luke A. Barnes.

Library of Congress Cataloging-in-Publication Data

Names: Hossenfelder, Sabine, 1976– author.
Title: Existential physics : a scientist's guide to life's biggest questions /
Sabine Hossenfelder.
Description: [New York, New York] : Viking, [2022] | Includes
bibliographical references and index.
Identifiers: LCCN 2021046360 (print) | LCCN 2021046361 (ebook) |
ISBN 9781984879455 (hardcover) | ISBN 9781984879462 (ebook)
Subjects: LCSH: Physics—Philosophy. | Cosmology. | Quantum theory. |
Meaning (Philosophy)
Classification: LCC QC6 .H656 2022 (print) | LCC QC6 (ebook) |
DDC 530.01—dc23/eng20220228
LC record available at https://lccn.loc.gov/2021046360
LC ebook record available at https://lccn.loc.gov/2021046361

Printed in the United States of America
5   7   9   10   8   6   4

Designed by Amanda Dewey

To Stefan

It is far better to grasp the Universe as it really is than to persist in delusion, however satisfying and reassuring.

—*Carl Sagan*

CONTENTS

PREFACE

"Can I ask you something?" a young man inquired after learning that I am a physicist. "About quantum mechanics," he added, shyly. I was all ready to debate the measurement postulate and the pitfalls of multipartite entanglement, but I was not prepared for the question that followed: "A shaman told me that my grandmother is still alive. Because of quantum mechanics. She is just not alive here and now. Is this right?"

As you can tell, I am still thinking about this. The brief answer is, it's not totally wrong. The long answer will follow in chapter 1, but before I get to the quantum mechanics of deceased grandmothers, I want to tell you why I'm writing this book.

During more than a decade in public outreach, I noticed that physicists are really good at answering questions, but really bad at explaining why anyone should care about their answers. In some research areas, a study's purpose reveals itself, eventually, in a marketable product. But in the foundations of physics—where I do most of my research—the primary product is knowledge. And all too often, my colleagues and I present this knowledge in ways so abstract that no one understands why we looked for it in the first place.

Not that this is specific to physics. The disconnect between experts and non-experts is so widespread that the sociologist Steve Fuller claims that academics use incomprehensible terminology to keep insights sparse and thereby more valuable. As the American journalist and Pulitzer Prize winner Nicholas Kristof complained, academics encode "insights into turgid prose" and "as a double protection against public consumption, this gobbledygook is then sometimes hidden in obscure journals."

Case in point: People don't care much whether quantum mechanics is predictable; they want to know whether their own behavior is predictable. They don't care much whether black holes destroy information; they want to know what will happen to the collected information of human civilization. They don't care much whether galactic filaments resemble neuronal networks; they want to know if the universe can think. People are people. Who'd have thought?

Of course, I want to know these things too. But somewhere along my path through academia I learned to avoid asking such questions, not to mention answering them. After all, I'm just a physicist. I'm not competent to speak about consciousness and human behavior and such.

Nevertheless, the young man's question drove home to me that physicists *do* know some things, if not about consciousness itself, then about the physical laws that everything in the universe—including you and I and your grandmother—must respect. Not all ideas about life and death and the origin of human existence are compatible with the foundations of physics. That's knowledge we should not hide in obscure journals using incomprehensible prose.

It's not just that this knowledge is worth sharing; keeping it to ourselves has consequences. If physicists don't step forward and explain what physics says about the human condition, others will jump at the opportunity and abuse our cryptic terminology for the promotion of pseudoscience. It's not a coincidence that quantum entanglement and vacuum energy are go-to explanations of alternative healers, spiritual media, and snake oil sellers. Unless you have a PhD in physics, it's hard to tell our gobbledygook from any other.

However, my aim here is not merely to expose pseudoscience for what it is. I also want to convey that some spiritual ideas are perfectly compatible with modern physics, and others are, indeed, supported by it. And why not? That physics has something to say about our connection to the universe is not so surprising. Science and religion have the same roots, and still today they tackle some of the same questions: Where do we come from? Where do we go to? How much can we know?

When it comes to these questions, physicists have learned a lot in the past century. Their progress makes clear that the limits of science are not fixed; they move as we learn more about the world. Correspondingly, some belief-based explanations that once aided sense-making and gave comfort we now know to be just wrong. The idea, for example, that certain objects are alive because they are endowed with a special substance (Henri Bergson's "élan vital") was entirely compatible with scientific fact two hundred years ago. But it no longer is.

In the foundations of physics today, we deal with the laws of nature that operate on the most fundamental level. Here, too, the knowledge we gained in the past hundred years is now replacing old, belief-based explanations. One of these old explanations is the idea that consciousness requires something more than the interaction of many particles, some kind of magic fairy dust, basically, that endows certain objects with special properties. Like the élan vital, this is an outdated and useless idea that explains nothing. I will get to this in chapter 4, and in chapter 6 I'll discuss the consequences this has for the existence of free will. Another idea ready for retirement is the belief that our universe is especially suited to the presence of life, the focus of chapter 7.

However, demarcating the current limits of science doesn't only destroy illusions; it also helps us recognize which beliefs are still compatible with scientific fact. Such beliefs should maybe not be called *un*scientific but rather *a*scientific, as Tim Palmer (whom we'll meet later) aptly remarked: science says nothing about them. One such belief is the origin of our universe. Not only can we not currently explain it, but also it is questionable whether we will ever be able to explain it. It may be one of the ways that science is fundamentally limited. At least that's what I currently believe. The idea that the universe itself is conscious, I have found to my own surprise, is difficult to rule out entirely (chapter 8). And the jury is still out on whether or not human behavior is predictable (chapter 9).

In brief, this is a book about the big questions that modern physics raises, from the question whether the present moment differs from the past, to the idea that each elementary particle may contain a universe, to the worry that the laws of nature determine our decisions. I cannot, of course, offer final answers. But I want to tell you how much scientists currently know, and also where science crosses over into mere speculation.

I will mostly stick with established theories of nature that are backed up by evidence. All of what I am going to say, therefore, should come with the preamble "as far as we currently know," meaning that further scientific progress might lead to revision. In some cases, the answer to a question depends on properties of natural laws that we do not yet fully understand, like quantum measurements or the nature of space-time singularities. If so, I will point out how future research could help answer the question. Because I don't want you to hear just my own opinion, I have added a few interviews. And at the end of the book, you'll find a brief glossary with definitions of the most important technical terms. Terms in the glossary are marked bold when they first appear in the text hereafter.

Existential Physics is for those who have not forgotten to ask the big questions and are not afraid of the answers.

A WARNING

I want you to know what you are getting yourself into, so let me put my cards on the table up front. I am both agnostic and a heathen. I have never been part of an organized religion and never felt the desire to join one. Still, I am not opposed to religious belief. Science has limits, and yet humanity has always sought meaning beyond those limits. Some do it by studying holy scripture, some meditate, some dig philosophy, some smoke funny things. That's all fine with me, really. Provided that—and here's the crux—your search for meaning respects scientific fact.

If your belief conflicts with empirically confirmed knowledge, then you are not seeking meaning; you are delusional. Maybe you'd rather hold on to your delusions. Trust me; I am sympathetic to that—but then this book is not for you. In the coming chapters, we will talk about free will, afterlife, and the ultimate search for meaning. It won't always be easy. I myself have struggled with some of the consequences of what I know to be well-confirmed natural laws, and I suspect some of you will find it equally difficult.

You may think I exaggerate to make dry physics sound more exciting. Look, we all know I want this book to sell, so why pretend otherwise? But the main reason I issue this warning is that I am sincerely worried that this book may negatively affect some readers' mental health. Occasionally someone contacts me, writing that they came across one of my essays, and now they don't know how to go on with their life. They seem genuinely disturbed. What sense does life make without free will? What's the point of human existence if it's just a random fluke? How can you not freak out knowing that the universe might blink out any moment?

Indeed, some scientific facts are hard to stomach and, worse, there's no psychologist who'll be able to help. I know this because I've tried. But hang on. If you think it through, science gives more than it takes. In the end, I hope you will find comfort in knowing that you do not need to silence rational thought to make space for hope, belief, and faith.

EXISTENTIAL PHYSICS

Chapter 1

DOES THE PAST STILL EXIST?

Now and Never

Time is money. It's also running out. Unless, possibly, it's on your side. Time flies. Time is up. We talk about time . . . all the time. And yet time has remained one of the most difficult-to-grasp properties of nature.

It didn't help that Albert Einstein made it personal. Before Einstein, everybody's time passed at the same rate. Post-Einstein, we know that the passage of time depends on how much we move around. And while the numerical value we assign to each moment—say 2:14 p.m.—is a matter of convention and measurement accuracy, in pre-Einstein days, we believed that *your* now was the same as *my* now; it was a universal now, a cosmic ticking of an invisible clock that marked the present moment as special. Since Einstein, *now* is merely a convenient word that we use to describe our experience. The present moment is no longer of fundamental significance because, according to Einstein, the past and the future are as real as the present.

This doesn't match with my experience and probably doesn't

match with yours either. But human experience is not a good guide to the fundamental laws of nature. Our perception of time is shaped by circadian rhythms and our brain's ability to store and access memories. This ability is arguably good for many things, but to disentangle the physics of time from our perception of it, it is better to look at simple systems, like swinging pendulums, orbiting planets, or light that reaches us from distant stars. It is from observations on such simple systems that we can reliably infer the physical nature of time without getting bogged down by the often inaccurate interpretation that our senses add to the physics.

A hundred years' worth of observation have confirmed that time has the properties Einstein conjectured at the beginning of the twentieth century. According to Einstein, time is a dimension, and it joins with the three dimensions of space to one common entity: a four-dimensional space-time. The idea of combining space and time to space-time goes back to the mathematician Hermann Minkowski, but Einstein was the one to fully grasp the physical consequences, which he summarized in his theory of special relativity.

The word *relativity* in *special relativity* means there is no absolute rest; you can merely be at rest relative to something. For example, you are now probably at rest relative to this book; it's moving neither away from nor toward you. But if you throw it into a corner, there are two ways of describing the situation: the book moves at some velocity relative to you and the rest of planet Earth, or you and the rest of the planet move relative to the book. According to Einstein, both are equivalent ways to describe the physics and should give the same prediction—that's what the word *relativity* stands for. The *special* just says that this theory doesn't include gravity. Gravity was included only later, in Einstein's theory of **general relativity**.

The idea that we should be able to describe physical phenomena the same way regardless of how we move in Einstein's four-dimensional space-time sounds rather innocuous, but it has a host of counterintuitive consequences that have entirely changed our conception of time.

o   o   o

In our usual three-dimensional space, we can assign coordinates to any location using three numbers. We could, for example, use the distance to your front door in the directions east-west, north-south, and up-down. If time is a dimension, we just add a fourth coordinate, let's say the time that has passed at your front door since 7:00 a.m. We then call the complete coordinates an *event*. For example, the space-time event at 3 meters east, 12 meters north, 3 meters up, and 10 hours might be your balcony at 5:00 p.m.

This choice of coordinates is arbitrary. There are many different ways to put coordinate labels on space-time, and Einstein said these labels shouldn't matter. The time that actually passes for an object can't depend on what coordinates we chose. And he showed that this invariant, internal time—*proper time*, as physicists call it—is the length of a curve in space-time.

Suppose you go on a road trip from Los Angeles to Toronto. What matters to you is not the straight-line coordinate distance between these points, about 2,200 miles, but the distance on highways and streets, which is more like 2,500 miles. It's similar in space-time. What matters is the length of the trip, not the coordinate distance. But there's an important difference: in space-time, the longer the curve between two events, the *less* time passes on it.

How do you make a curve between two space-time events longer? By changing your velocity. The more you accelerate, the slower your proper time will pass. This effect is called *time dilation*. And, yes, in principle, this means if you run in a circle, you'll age more slowly. But it's a tiny effect, and I can't recommend it as an antiaging strategy. By the way, this is also why time passes more slowly near a black hole than far away from one. That's because, according to Einstein's principle of equivalence, a strong gravitational field has the same effect as a fast acceleration.

What does this mean? Imagine I have two identical clocks; I hand you one, and then you go your way and I go mine. In pre-Einstein days, we'd have thought that whenever we met again, these clocks would show exactly the same time—this is what it means for time to be a universal parameter. But post-Einstein, we know this isn't right. How much time passes on your clock depends on how much and how fast you move.

How do we know this is correct? Well, we can measure it. It would lead us too far off topic to go into detail about which observations have confirmed Einstein's theories, but I will leave you recommendations for further reading in the endnotes. To move on, let me just sum it up by saying that the hypothesis that the passage of time depends on how you move is supported by a large and solid body of evidence.

I have been speaking of clocks for illustration, but the fact that acceleration slows time down has nothing in particular to do with the devices we call clocks; it happens for any object. Whether it's combustion cycles, nuclear decay, sand running through an hourglass, or heartbeats, each process has its own individual passage of time. But the differences between individual times are normally minuscule, which is why we don't notice them in everyday life. They become noticeable, however, when we keep track of time very precisely, which we do, for example, in satellites that are part of the global positioning system (GPS).

The GPS, which your phone's navigation system most likely uses, allows a receiver—like your phone—to calculate its position from signals of several satellites that orbit Earth. Because time is not universal, time on these satellites passes subtly differently compared with how it passes on Earth, both because of the satellites' motion relative to the surface of Earth and because of the weaker gravitational field that the satellites experience in their orbits. The software on your phone needs to take this into account to correctly infer its location, because the different passage of time on the satellites oh-so-slightly distorts the signals. It's a small effect, all right, but it's not philosophy; it's physically real.

o o o

The fact that the passage of time isn't universal is pretty mind-bending already, but there's more. Because the speed of light is very fast but finite, it takes time for light to reach us, so, strictly speaking, we always see things as they looked a little bit earlier. Again, though, we don't normally notice this in everyday life. Light travels so fast that it doesn't matter on the short distances we see on Earth. For example, if you look up and watch the clouds, you actually see the clouds the way they looked a millionth of a second ago. That doesn't really make a big difference, does it? We see the Sun as it looked eight minutes ago, but because the Sun doesn't normally change all that much in a few minutes, light's travel time doesn't make a big difference. If you look at the North Star, you see it as it looked 434 years ago. But, yeah, you may say, so what?

It is tempting to attribute this time lag between the moment something happens and our observation of it as a limitation of perception, but it has far-reaching consequences. Once again, the issue is that the passage of time is not universal. If you ask what happened "at the same time" elsewhere—for example, just exactly what you were doing when the Sun emitted the light you see now—there is no meaningful answer to the question.

This problem is known as the *relativity of simultaneity*, and it was well illustrated by Einstein himself. To see how this comes about, it helps to make a few drawings of space-time. It's hard to draw four dimensions, so I hope you will excuse me if I use only one dimension of space and one dimension of time. An object that doesn't move relative to the chosen coordinate system is described by a vertical straight line in this diagram (figure 1). These coordinates are also referred to as the *rest frame* of the object. An object moving at constant velocity makes a straight line tilted at an angle. By convention, physicists use a 45-degree angle for the speed of light. The speed of light is the same for all observers, and because it can't be exceeded, physical objects have to move on lines tilted less than 45 degrees.

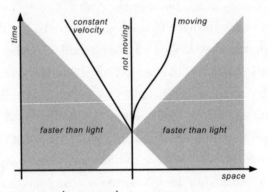

Figure 1: How space-time diagrams work.

Einstein now argued as follows. Let's say you want to construct a notion of simultaneity by using pulses of laser beams that bounce off mirrors that are at rest relative to you.[a] You send one pulse to the right and one to the left and shift your position between the mirrors until the pulses return to you at the same moment (see figure 2a). Then you know you are exactly in the middle and the laser beams hit both mirrors at the same moment.

---

[a] I myself used to be perplexed about what makes lasers so special that they constantly appear in books about space-time. The answer is "Nothing really." It's just that because we know laser light moves at the speed of light (duh) and doesn't spread (much), lasers are particularly handy to illustrate the relation between space and time.

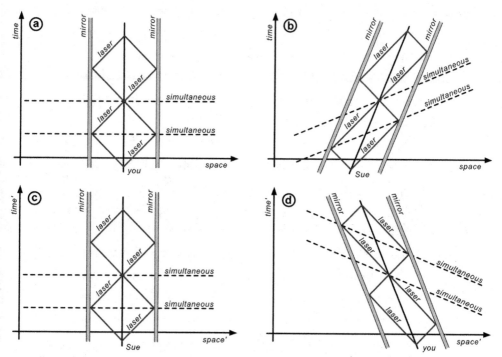

Figure 2: Space-time diagrams for construction of simultaneous events. Top left (a): You in your rest frame with coordinates labeled space and time. Top right (b): Sue in your rest frame. Bottom left (c): Sue in her rest frame with coordinates labeled space' and time'. Bottom right (d): You in Sue's rest frame.

Once you have done that, you know at exactly which moment in your own time the laser pulse will hit both mirrors, even though you can't see it because the light from those events hasn't yet reached you. You could look at your clock and say, "Now!" This way, you have constructed a notion of simultaneity that, in principle, could span the whole universe. In practice you may not have the patience to wait ten billion years for the laser pulse to return, but that's theoretical physics for you.

Now imagine that your friend Sue moves relative to you and tries to do the same thing (figure 2b). Let's say she moves from left to right. Sue, too, uses two mirrors, one to her right and one to her left, and the mirrors move along with her at the same velocity—hence, the mirrors are in rest relative to Sue, like your mirrors are relative to you. Like you, she sends laser pulses in both directions and positions herself so the pulses come back to her from both sides at the same moment. Like you, she then knows that the pulses hit the two mirrors at the same moment, and she can calculate just which moment that corresponds to on her own clock.

The trouble is, she gets a different result than you do. Two events that Sue thinks happen at the same time would not happen at the same time according to you. That's because from your perspective she is moving toward one of the mirrors and away from the other. To you it seems that the time it takes the pulse to reach the mirror on her left is shorter than the time it takes for the other pulse to catch up with the mirror on her right. It's just that Sue doesn't notice, because on the pulses' return paths from the mirrors, the opposite happens. The pulse from the mirror to Sue's right takes longer to catch up with her, while the pulse from the mirror on her left arrives faster.

You would claim that Sue is making a mistake, but according to Sue, *you* are making the mistake because, to her, you are the one who is moving. She would say that actually *your* laser pulses do not hit your mirrors at the same time (figures 2c and 2d).

Who is right? Neither of you. This example shows that in special relativity the statement that two events happened at the same time is meaningless.

It's worth stressing that this argument works only because light doesn't need a medium to travel in, and the speed of light (in vacuum) is the same for all observers. This argument does not work with sound waves, for example (or any other signal that isn't light in vacuum), because then the speed of the signal really will not be the same for all observers; it will instead depend on the medium it's traveling in. In that case, one of you would be objectively right and the other one wrong. That your notion of now might not be the same as mine is an insight we owe to Albert Einstein.

o   o   o

We just established that two observers who move relative to each other don't agree on what it means for two events to happen at the same time. That isn't only odd, but it entirely erodes our intuitive notion of reality.

To see this, suppose you have two events that are not in causal contact with each other, which means you cannot send a signal from one to the other, not even at the speed of light. Diagrammatically, "not in causal contact" just means if you draw a straight line through the two events, the angle between the line and the horizontal is less than 45 degrees. But look at figure 2b again. For two events that are not in causal contact, you can always imagine an observer for whom everything on this straight line is simultaneous. You just need to choose the observer's velocity so the return points of the laser pulses are on the line. But if any two points that are not causally connected happen at the same time for someone, then every event is "now" for someone.

To illustrate the latter step, let us say the one event is your birth and the other event is a supernova explosion (see figure 3). The explosion is causally disconnected from your birth, which means the light from it hadn't reached Earth at the time you were born. You can then imagine that your friend Sue, the space traveler, sees these events at the same time, so they happened simultaneously according to her.

Suppose further that by the time you die the light from the supernova still hasn't reached Earth. Then your friend Paul could find a way to travel in the middle between you and the supernova so he would see your death and the supernova at the same time. They both

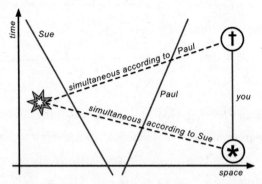

Figure 3: *Any two causally disconnected events are simultaneous for some observers. If all observers' experiences are equally valid, then all events exist the same way, regardless of when or where they are.*

happened simultaneously according to Paul. I swear that's it for introducing imaginary friends on spaceships!

We can then put together everything we learned. I believe most of us would say the clouds exist now, even though we can see them only as they were a fraction of a second ago. For this, we use our own, personal notion of simultaneity that depends on how we move through space-time—that is, usually much below the speed of light and on the surface of our planet. Therefore, we all pretty much mean the same thing by "now," and it doesn't normally cause confusion.

However, all notions of "now" for observers who move elsewhere and potentially close to the speed of light—like Sue and Paul—are equally valid, and in principle they span the entire universe. And because there could be some observer according to whom your birth and the supernova explosion happen simultaneously, the supernova exists at your birth according to your own notion of existence. Therefore, because there could be another observer according to whom the explosion happens together with your death, your death exists at your birth.

You can advance this argument for any two events anywhere in the universe at any time and arrive at the same conclusion: the physics of

Einstein's special relativity does not allow us to constrain existence to merely a moment that we call "now." Once you agree that *anything* exists now elsewhere, even though you see it only later, you are forced to accept that *everything* in the universe exists now.

This perplexing consequence of special relativity has been dubbed the *block universe* by physicists. In this block universe, the future, present, and past exist in the same way; it's just that we do not experience them the same way. And if all times exist similarly, then all our past selves—and grandparents—are alive the same way our present selves are. They are all there, in our four-dimensional space-time, have always been there, and will always be there. To sum it up in the words of the British comedian John Lloyd, "Time is a bit like a landscape. Just because you're not in New York doesn't mean it's not there."

More than a century has passed since Einstein put forward his theories of special and general relativity. But here we are today, still struggling to understand what it really means. It sounds crazy, but the idea that the past and future exist in the same way as the present is compatible with all we currently know.

Eternal Information

The notion that the present moment has no special relevance can be seen another way. All successful theories in the **foundations of physics** require two ingredients: (1) information about what it is that you want to describe at one moment in time, called the **initial condition**, and (2) a prescription, called an **evolution law**, for how to calculate from this **initial state** what happens at another moment of time.

I want to caution you that the word *evolution* here has nothing to do with Charles Darwin; it merely means that the law tells us how a system evolves—that is, changes in time. For example, if you know the place and velocity of a meteorite entering Earth's atmosphere (initial condition), applying the evolution law allows you to calculate its place of impact. And because we are introducing terminology already, the technical expression for "that which you want to describe" is *system*. No, seriously. While *system* has a rather specific meaning in other disciplines, among physicists it can mean anything and everything. That's very convenient, so it's also how I will use the word.

Thus, when we want to make a prediction, we take the state of a system at one time, and then we use the evolution law to calculate from this one time what the system will do at any other time. But we can do this in either direction of time. The laws, as we say, are **time-reversible**. They can be run forward and backward, like a movie.

In our everyday experience, forward in time looks very different from backward in time. We see eggs breaking but not unbreaking, logs burning but not unburning, people aging but not getting any younger. I have dedicated the entire chapter 3 to the question of why forward in time looks different from backward in time. But for this chapter, I will put aside the question why time seems to have a preferred direction and instead look just at the consequences of the time-reversibility of the laws.

Time-reversibility does not mean that both directions in time look the same; that would be called *time-reversal invariance*. Time-reversibility merely means that, given the entire information at one moment, we can calculate what happened at any moment before that and what will happen at any moment after that.

The idea that all events in the future can in principle be calculated from any earlier time is called **determinism**. Prior to the discovery of **quantum mechanics**, the then-known laws of nature were deterministic. In 1814, the French scientist and philosopher Pierre-Simon Laplace conjured up a fictional, omniscient being to illustrate the consequences.

We ought then to regard the present state of the universe as the effect of its anterior state and as the cause of the one which is to follow. Given for one instant an intelligence which could comprehend all the forces by which nature is animated and the respective situation of the beings who compose it—an intelligence sufficiently vast to submit these data to analysis—it would embrace in the same formula the movements of the greatest bodies of the universe and those of the lightest atom; for it, nothing would be uncertain and the future, as the past, would be present to its eyes.

This omniscient being, *Laplace's demon*, is an ideal. In practice, of course, no one has all the information necessary to predict the future with certainty—we aren't omniscient. But I am here not concerned with what calculation can be done in practice; I want to look at what the fundamental laws and their properties tell us about the nature of reality.

Now, a time-reversible law is also deterministic, but the opposite is not necessarily true. Imagine a video game that can't be won. You watch recordings of gamers playing but ultimately always losing the game. Inevitably, the recording will end with the same screen saying, GAME OVER. This means if you see only the end screen, you can't tell what happened previously. The outcome is determined, but not time-reversible. A time-reversible law, in contrast, results in a unique relationship between any two moments of time. For the example of the video game, this would mean that the final screen contains enough details for you to figure out exactly which moves led to this outcome.

The currently known fundamental laws of nature are both time-reversible and deterministic, with the exception of two processes that I will discuss in the next section. That the future is fixed by the present in this way seems to severely constrain our ability to make decisions. We will talk about what this means for free will in chapter 6. For now, I want to focus on the brighter side of time-reversal invariance, which is that the universe keeps a faithful record of the information about all you have ever said, thought, and done.

I use the word *information* here loosely to refer to all numbers you need to put into the evolution law to be able to make a prediction with it. Information, hence, is merely all the details you need in order to completely specify the initial state of the system at one particular time. In other areas of physics, information has properties beyond that, but that's the way I will use the term here.

The evolution law maps the initial state at any one time to the state at any other time, so it really just tells us how matter in the universe and space-time reconfigures. We start with particles in one arrangement, we apply the equation to it, and we get another arrangement. The information in these arrangements is completely maintained. To recover an earlier state, all you need to do is apply the evolution law and run it backward. In practice, this is unfeasible. But in principle, information—including every oh-so-minute detail about your identity—cannot be destroyed.

o o o

Let us then talk about the two exceptions to time-reversibility: the measurement in quantum mechanics, and the evaporation of black holes.

Quantum mechanics has a time-reversible evolution law (the Schrödinger equation) for a mathematical object called the *wave function*. The wave function is usually denoted by Ψ (the Greek capital letter psi) and it describes whatever it is you want to observe (the "system" again). From the wave function, we compute probabilities for measurement outcomes, but the wave function itself is not observable.

To see how this works, consider the following example. Suppose we use quantum mechanics to calculate the probability for a particle to be measured at a particular place. To detect the particle, we use a luminous screen that emits a flash where the particle hits it. Let us say our calculation predicts there's a 50 percent chance we will find the particle on the left side of the screen and a 50 percent chance we'll find it on the right side. According to quantum mechanics, this probabilistic prediction is all there is to say. It is probabilistic not because we are missing information. There just isn't any more information. The wave function is the full description of the particle—that's what it means for the theory to be fundamental.

However, the moment we actually measure the particle, we know for sure whether it's on one side of the screen or the other. This means we have to update the wave function from 50:50 to either 100:0 or 0:100, depending on which side of the screen we saw the particle on. This update is sometimes also called the *reduction* or the *collapse* of the wave function. I find the word *collapse* misleading because it suggests a physical process that quantum mechanics doesn't contain, so I will stick with *update* or *reduction*. Without the update, quantum mechanics just does not describe what we observe.

"But what is a measurement?" you may ask. Yes, good question. This certainly bothered physicists a lot in the early days of quantum mechanics. By now this question has, luckily, largely been answered. A measurement is any interaction that is sufficiently strong or frequent to destroy the quantum behavior of a system. Only what it takes to destroy quantum behavior can be (and, for many examples, has been) calculated.

Most important, these calculations show that a measurement in quantum mechanics does not require a conscious observer. In fact, it doesn't even require a measurement apparatus. Even tiny interactions with air molecules or light can destroy quantum effects so that we have to update the wave function. Of course, in this case, speaking of a measurement is quite the abuse of language, but physically there isn't any difference between interactions with a man-made apparatus and interactions with a naturally present environment. And because in everyday life we can't ever get rid of the environment, we don't normally see quantum effects, like dead-and-alive cats, with our own eyes. Quantum behavior just gets destroyed too easily.

This is also why you shouldn't listen to anyone who claims that quantum leaps allow you to think your way out of illness or that you can improve your life by drawing energy from quantum fluctuations and so on. This isn't just off-the-mainstream science; it's incompatible with evidence. Under normal circumstances, quantum effects don't play a role beyond the size of molecules. That they're difficult to maintain and measure is the very reason physicists like doing experiments at temperatures near absolute zero, preferably in vacuum.

We understand fairly well what constitutes a measurement, but the fact that we need to update the wave function upon measurement makes quantum mechanics both indeterministic and time-irreversible. It is indeterministic because we cannot predict what we will actually measure; we can predict only the probability of measuring something. And it is not time-reversible, because once we have measured the particle, we cannot infer what the wave function was prior to measurement. Suppose you measure the particle on the left side of your screen. Then you cannot tell whether the wave function previously said the particle should be there with 50 percent probability or with a mere 1 percent probability. There are many different initial states for the wave function that will result in the same measurement outcome. This means the measurement in quantum mechanics destroys information for good.

However, if you know one thing about quantum mechanics, it's that its physical interpretation has remained highly controversial. In 1964, more than half a century after the theory was established, Richard Feynman told his students, "I can safely say that nobody understands quantum mechanics." After another half century, in 2019, the physicist Sean Carroll wrote that "even physicists don't understand quantum mechanics."

Indeed, the fact that the wave function can't itself be observed is a dilemma that has kept physicists and philosophers up at night for the better part of a century, but we don't need to go through the whole discussion here. If you want to know more about the interpretations of quantum mechanics, please have a look at my reading suggestions in the endnotes. Let me just sum it up by saying that if you don't believe the measurement update is fundamentally correct, that's currently a scientifically valid position to hold. I myself think it's likely the measurement update will one day be replaced by a physical process in an underlying theory, and it might come out to be both deterministic and time-reversible again.

I should add that in one of the currently most popular interpretations of quantum mechanics—the many-worlds interpretation— the measurement update does not happen at all, and the evolution of the universe just remains time-reversible. I am not a big fan of the many-worlds interpretation for reasons I will lay out in chapter 5, but to give you an accurate impression of the current status of research, the many-worlds interpretation is another reason that believing in time-reversibility is presently compatible with scientific knowledge.

This brings us to the other exception to time-reversibility: the evaporation of black holes. Black holes are regions where space-time bends so strongly that light is forced to go around in circles and can't escape. The surface within which light gets trapped is called the *horizon* of the black hole; in the simplest case, the horizon has the shape of a sphere. Because nothing can move faster than light, black holes will trap everything that crosses the horizon. If something happens to fall in—an atom, a book, a spaceship—it can't get back out, ever. Once inside the black hole, it's eternally disconnected from the rest of the universe.

However, just because something is out of sight doesn't mean it has stopped existing. If I put a book into a box, I can also no longer see it, but that doesn't destroy the information in the book. The mere presence of a black hole horizon is therefore not a problem for the preservation of information. It certainly is a problem for the *accessibility* of the information, but if black holes just continued to store information indefinitely, that'd be entirely unproblematic.

And that was the status until, in 1974, Stephen Hawking showed that black holes don't live forever. Because of quantum fluctuations, space-time around the black hole horizon becomes unstable. In this region, previously empty space decays into particles, primarily into photons (the particles of light) and particles of tiny mass called *neutrinos*. This creates a steady stream, called *Hawking radiation*, that carries energy away from the horizon. The black hole evaporates, and because energy is conserved, the black hole shrinks.

However, because Hawking radiation does not come from inside the black hole, it cannot contain information about what originally formed the black hole or what fell in later. Remember that what's inside the black hole is disconnected from the outside. The radiation does carry a few bits of information. For example, if you catch it all, you can infer the total mass and angular momentum of the black hole. But the radiation does not carry remotely enough information to encode all details of what vanished behind the horizon. Therefore, when the black hole has entirely evaporated, and the only thing that's left is the Hawking radiation, you have no way to figure out what the initial state was. Was it once a white dwarf or a neutron star? Did it eat up a small moon, or a hydrogen cloud, or an unlucky space traveler? What were the space traveler's final words? You can't tell. The evaporation of a black hole is thus time-irreversible: there are many different initial states that result in the same final state.

This sounds superficially similar to the measurement problem, but there is an important difference. The destruction of information in black hole evaporation happens even before one measures the radiation. That's a big problem because it means black hole evaporation is incompatible even with the evolution law of quantum theory. It is for this reason that most physicists currently think something is wrong about Hawking's conclusion that black holes destroy information.

Hawking himself, in his later years, changed his mind and became convinced that black holes do not destroy information. The most obvious shortcoming of Hawking's 1974 calculation is that it does not include the quantum properties of gravity. It can't, because we don't have a theory for that. If we had such a theory, and if we included its effects, maybe that would restore time-reversibility in black hole evaporation. A lot of physicists currently think it would.

o o o

In summary, other than quantum measurements and black hole evaporation, both of which are controversial cases, information can't be destroyed. I find much solace in this knowledge when I misplace my car keys. More seriously, of course, once your grandmother dies, information about her—her unique way of navigating life, her wisdom, her kindness, her sense of humor—becomes, in practice, irretrievable. It disperses quickly into forms we can no longer communicate with and that may no longer allow an experience of self-awareness. Nevertheless, if you trust our mathematics, the information is still there, somewhere, somehow, spread out over the universe, but preserved forever. It might sound crazy, but it's compatible with all we currently know.

Transcendent Math

My arguments in this chapter so far relied on analyzing mathematical properties of the laws of nature, which is a method that itself warrants further inspection. It is a curious fact that mathematics is "unreasonably effective" in the natural sciences, as Eugene Wigner put it so unforgettably. Indeed, mathematics has worked incredibly well for physicists; the proof is in front of your eyes. Whether you read this book on a screen or laser-printed on paper, it was brought to you by physicists who dug deep into the math of quantum mechanics on which modern technology relies. You may not know the math, you may not understand it, or you may not like it, but there is no doubt that it works.

And yet physics isn't math. Physics is a science and as such has the purpose of describing observations of natural phenomena. Yes, we use mathematics in physics, and plenty of that, as I'm sure you have noticed. But we do this not because we know the world is truly mathematics. It may be mathematics—this possibility is known as *Platonism*, but Platonism is a philosophical position, not a scientific one. All we can tell from observations is that math is *useful* to describe the world. That the world *is* math—rather than just being described by math—is an additional assumption. And because this additional assumption is unnecessary to explain what we observe, it's not scientific.

However, the belief that reality is math is deeply ingrained into the thinking of many physicists who treat mathematics as a timeless realm of truth that we reside in. It is common for textbooks and papers to state that space-time *is* a particular mathematical structure, and that particles *are* certain mathematical objects. Physicists may not consciously subscribe to the idea that math is real and when asked will deny it, but in practice they do not distinguish the two. This conflation has consequences, for they sometimes erroneously come to think their math reveals more about reality than it possibly can.

This is most obvious in Max Tegmark's idea of the "mathematical universe." According to Tegmark, all of mathematics is real and it's all equally real, not just the math that describes our observations, but literally any math: Euler's number, the zeros of the Riemann zeta function, pseudometric non-Hausdorff manifolds, moduli spaces of p-adic Galois representations—all as real as your big toe.

You may find that a little hard to swallow. But however you feel about it, it's not wrong; it's just not scientific. We clearly don't need all of mathematics to describe our observations—the universe is one way and not any other, so describing it requires only very specific math. And scientific hypotheses should not have superfluous assumptions, for that would allow adding statements like "and God made it." Postulating that all math is real is such an unscientific, superfluous assumption—it doesn't help us describe nature any better. But just because there's a lot of math that we don't need doesn't mean it does not exist either. Postulating that it doesn't exist is also superfluous to describing our observations. So, as with God, science can't say anything about whether or not all that math exists.

Frankly, I think Tegmark came up with the mathematical universe only to make sure everyone knows he is a seriously weird fellow. He was probably successful at that, but whatever his motivation, I will admit that to me the thought that reality is just a manifestation of absolute mathematical truths is a comforting belief. If it were so, then at least the world would make sense; it's just that we don't know or don't understand the mathematics to make sense of it.

However, while I find it comforting to think that reality is mathematics, I can't actually get myself to believe it. It strikes me as presumptuous to think that humans have already discovered the language in which nature speaks, basically on the first try and right after we appeared on the surface of the planet. Who is to say there may not be a better way to understand our universe than mathematics, one that may take us a million years to figure out? Call it the *principle of finite imagination*: Just because we can't currently think of a better explanation doesn't mean there isn't one. Just because we don't yet know a better way to describe natural phenomena than mathematics doesn't mean there isn't one.

So if you want to believe that the past exists because it's math and all of math exists, that is up to you. The arguments in the previous sections of this chapter do not depend on whether you believe in the reality of math. However, they implicitly assume that mathematics itself is timeless, that mathematical truth is eternal, and that logic doesn't change. This is an assumption that cannot be proved, because what would you prove it true with? It's one of the usually unstated articles of faith that our scientific inquiry is based on.

>> THE BRIEF ANSWER

According to the currently established laws of nature, the future, the present, and the past all exist in the same way. That's because, regardless of exactly what you mean by *exist*, there is nothing in these laws that distinguishes one moment of time from any other. The past, therefore, exists in just the same way as the present. While the situation is not entirely settled, it seems that the laws of nature preserve information entirely, so all the details that make up you and the story of your grandmother's life are immortal.

Chapter 2

HOW DID THE UNIVERSE BEGIN? HOW WILL IT END?

What Does It Mean to Explain Something?

Planet Earth formed around 4.5 billion years ago. The first primitive forms of life appeared about 4 billion years ago. Natural selection did the rest, giving rise to species increasingly better adapted to their environment. The evidence, as they say, is overwhelming.

Or is it? Imagine that planet Earth began its existence a mere six thousand years ago, with all fossil records in place and stones well weathered. From there on, however, evolution proceeded as scientists say. How would you prove this story wrong?

You couldn't.

I am sorry, but I told you it wouldn't be easy!

It is impossible to prove this story wrong, because of the way our current natural laws work. As we discussed in the previous chapter, they work by applying evolution laws to initial states, and we can apply those evolution laws both forward and backward in time. If we want to make a prediction for the path of a celestial object, we measure its present location and velocity and evolve it forward. If we want to know how the universe looked billions of years ago, we use our observations from the present time and then run the equations backward.

This method creates the following problem, however. If I take a present state, like the Earth in the year 2022, and apply an evolution law to it, then that will give me a past state in 3978 BCE. If I then take that past state and evolve it forward in time again, I will correctly get back to the year 2022. Trouble is, I can do that for *any* evolution law. There is always *some* state six thousand years ago that, together with the right evolution law, will correctly result in what we observe today.

Indeed, if I wanted to, I could suddenly switch to a different evolution law more than six thousand years in the past, to accommodate a creator, or the construction of a supercomputer that runs the cosmic simulation we all reside in, or really whatever I want. This is why, with natural laws like the ones we currently use, the idea that Earth was created by someone or something with everything in place is impossible to rule out.

Because such creation stories can't be falsified, we can't tell if they are false, but being false is not their problem. The problem with these stories is that they are bad scientific explanations.

The distinction between scientific and nonscientific explanations is central to this book, so it deserves a closer look. Science is about finding useful descriptions of the world; by *useful* I mean they allow us to make predictions for new experiments, or they quantitatively explain already existing observations. The simpler an explanation, the more useful it is. For a scientific theory, this explanatory power can be quantified in a variety of ways that come down to calculating how much input a theory needs to fit a set of data to a certain level of accuracy. Exactly how one quantifies explanatory power doesn't matter for our purposes. Let us just note that it can be done, and that it's something scientists actually do in some areas of science. Cosmology is one of the cases where this is done frequently.

In other areas of science, like biology or archaeology, mathematical models are not widely used now and therefore explanatory power usually can't be quantified. This is for a variety of reasons, but one is certainly that the observations themselves are often in qualitative, not quantitative, form. Now, a quantification of observations—made, say, by inventing a measure for the evil of war—doesn't necessarily bring more insights, so I'm not saying anything and everything needs to be cast into equations. But quantification can serve to remove doubts that conclusions were biased by human perception. This can be done, for example, to quantify the explanatory power of Darwinian evolution, by developing a mathematical measure of distance between fossils.

Scientific theories greatly simplify the stories we tell about the world, and that simplification embodies what we even mean by doing science. A good scientific theory is one that allows us to calculate the results of many observations from few assumptions. Quantum theory, to name just one, allows us to calculate the properties of the chemical elements. It is an extremely good scientific theory because it explains much from little. The belief that an omniscient being called God made the chemical elements is not a good scientific theory. You might say it is in some sense a simple explanation, and maybe you find it compelling. You may even find it necessary to make sense of your personal experience. However, the God hypothesis has no quantifiable explanatory power. You can't calculate anything from it. That doesn't make it wrong, but it does make it unscientific.

Saying that the world was created six thousand years ago with everything in place is unfalsifiable but also useless. It is quantifiably complicated: you need to put a lot of data into the initial condition. A much simpler, and thus scientifically better, explanation is that planet Earth is ages old and Darwinian evolution did its task.

Now that we know what it means to explain something in scientific terms, let us look at one of the cases where physicists currently struggle to find explanations: the beginning of our universe.

Modern Tales of Creation

In the beginning, superstrings created higher-dimensional membranes. That's one story I've been told, but there are many others. Some physicists believe the universe started with a bang, others think it was a bounce, yet again others bet on bubbles. Some say that everything began with a network. Some like the idea that it was a collision of sorts, or a timeless phase of absolute silence, or a gas of superstrings, or a five-dimensional black hole, or a new force of nature.

In the end, it doesn't matter—the outcome is the same: us, in a universe that looks like the one we see; that it doesn't matter which story you believe is a big warning sign. If this were science, we should have data to tell us which hypothesis is right, or at least an idea for obtaining the necessary data. But it's highly questionable that the data required to falsify any of these origin myths can be obtained, ever. These stories reach back in time so far that data are too sparse for astrophysicists to distinguish one tale from another, and this impasse might be impossible to overcome. For all we know, the beginning of our universe may remain hidden from us forever.

To see why I say this, I need to give you some background on how we develop theories for the early universe. We take all the data we can get, and then we look for a simple explanation. The more patterns in the data we can calculate with it, the better the explanation. For example, the current theory for the universe, the **concordance model**, is successful not just because, if fed with some initial condition, it gives us the present state. As noted earlier, this can always be done.

No, the relevant point is that the initial conditions are simple; they explain a lot from little.

The concordance model is an application of Einstein's theory of general relativity, according to which gravity is caused by the curvature of space-time. I will not go into this in detail here, because you don't need to know the details to follow along; you merely need to know that, according to general relativity, a universe filled with matter and energy will expand, and how fast it expands depends on the types and amounts of matter and energy in the universe. Hence, the concordance model basically keeps track of how much of which stuff is in the universe, from which we deduce the rate of expansion.

In physics we can run our models backward in time, and so, starting with the present state of the universe—expansion with matter clumped in galaxies—we can go back in time and deduce that the matter must have been squeezed together. It must once have been a hot and almost entirely smooth soup of elementary particles, called a *plasma*.

That the plasma was only *almost* entirely smooth is important. The plasma had small clumps in which the density was a tiny little bit larger than the average, and in other places, the density was a tiny little bit smaller. But gravity has the effect of drawing matter toward other matter. That is, gravity turns small clumps into bigger clumps. Incredible as it sounds, over the course of billions of years this makes the small irregularities in the plasma grow to entire galaxies. And the distribution of galaxies we observe today is then—through the evolution law—directly related to the distribution of the little clumps in the plasma in the early universe. Therefore, we can use the observations of galaxies today to infer, by running the evolution law backward, what the little clumps in the plasma must have looked like, how large they were, and how far apart from one another they were.

Moreover, the distribution of galaxies is not the only observation we can use to infer what the plasma must have looked like. That's because the spots in the plasma where the density was a little higher were also a little hotter, and the spots where the density was a little lower were a little cooler. Now, as long as the plasma is on average very dense, it is opaque, meaning that light will be swallowed almost immediately after being emitted. However, as the density of the plasma drops, elementary particles can stick together and form the first small atomic nuclei. After some hundred thousand years, there comes a moment—called *recombination*—when the plasma has cooled sufficiently so the atomic nuclei keep electrons bound to them.[a] After that, light is unlikely to be absorbed again. This light from recombination then streams freely through the expanding universe.

As the universe expands, the wavelength of the light stretches and so its vibrational frequency decreases. Because the frequency is proportional to the energy of the light, and the average energy determines the temperature, the temperature of the light drops with the expansion. This light is still around today, though at an extremely low temperature of 2.7 Kelvin (that is, 2.7 degrees Celsius above absolute zero); it makes up the *cosmic microwave background*. The name derives from the typical wavelength of the light, which is about 2 millimeters and falls into the microwave part of the electromagnetic spectrum.[b]

The temperature of the cosmic microwave background, however, isn't exactly the same in all directions of the sky. The *average* temperature is 2.7 Kelvin, but around that average there are small deviations of a few hundred-thousandths of a degree Kelvin. This means

---

[a]One of the countless mysteries of scientific terminology is why it's called *recombination* rather than just *combination*, given that it was possibly the first time they were ever combined. My best guess is that this term was borrowed from atomic physics, in which a plasma always first has to be heated before it can cool and recombine. The *re* probably stuck to *combined* just because the binding energy was too high to split it off.
[b]That's much shorter than the wavelengths used by microwave ovens, which are typically in the range of about 10 centimeters, or 4 inches.

that the light coming from some directions is a tiny little bit warmer and that from other directions is a tiny little bit colder. These temperature fluctuations in the cosmic microwave background also go back to the density fluctuations in the plasma in the early universe.

The important point now is that the initial conditions for the plasma in the early universe fit to both observations: the distribution of galaxies and the temperature variations in the cosmic microwave background. The concordance model of cosmology, therefore, is a simplification over just collecting the data: it explains why two different types of data fit together in a very specific way. While you can posit an initial condition to any evolution law so the result will agree with observations, you will in general have to put a lot of information into the initial condition to make the calculations come out just right to fit the observations. The concordance model, in contrast, does not need much information—neither in the dynamical law, nor in the initial condition—to explain several different observations. It makes things fit together. It has, in the words of the previous section, high explanatory power.

I have picked out two specific observations—the distribution of galaxies and the cosmic microwave background—to illustrate what I mean when I say the concordance model is a good explanation, but there are other observations that also fit it, such as the abundance of chemical elements and the way in which galaxies form. These observations strengthen the case for the concordance model.

The concordance model is considered a good scientific theory because it's simple yet it explains such a lot of data. The numerical values that currently fit best to the collected data tell us that only about 5 percent of the universe is made of the same stuff as we are, 26 percent is thinly distributed **dark matter**, which we can't see, and the remaining 69 percent is attributed to the **dark energy** of the **cosmological constant**.

How does the Big Bang fit into this model? The Big Bang refers to a hypothetical first moment in time when the universe began, so it would have happened before the hot-plasma phase we just discussed. If we go purely by the mathematics, then at the time of the Big Bang the matter in the universe must have been infinitely dense. An infinite density makes no physical sense, though, so it probably just signals that Einstein's theory of general relativity breaks down for very high densities. When physicists say "Big Bang," they therefore usually are not referring to the mathematical singularity but to whatever might replace the singularity in a better theory of space-time still to be found.[a]

The Big Bang, however, is not part of the concordance model. That's because we have no observation that tells us anything about what happened that far back in time. The problem is, when we run our equations backward in time, the density and temperature of the plasma continue to increase. Eventually, the plasma will be hotter and denser than what we have been able to produce in the world's most powerful particle colliders. And beyond the energy of those colliders, we no longer know what physical processes to expect. We have never tested this regime, and it doesn't occur in any other situation that we have observed. Even inside stars, temperatures and densities do not exceed the ones we have produced on Earth. The only naturally occurring event we know of that can reach higher densities is a star that collapses to a black hole. Alas, in this case, we can't observe what's going on, because the collapse is hidden behind the black hole horizon.

It's not a small gap in our knowledge. The energies at the Big Bang were at least fifteen orders of magnitude higher than the energies we

---

[a]Some physicists and science communicators use the term *Big Bang* to refer to times considerably later in the expansion of the universe. In this case, the Big Bang has nothing to do with the initial singularity. This has caused and continues to cause a lot of confusion, and I will not use the term in this sense here.

currently have reliable data about. Of course, we can speculate, and physicists have certainly speculated with abandon.

The straightforward speculation is to assume that nothing changes with the evolution equation of the concordance model, so we can just continue to roll it back in time, into the range for which we have no data. Just to give you a sense of what it means to extrapolate over fifteen orders of magnitude, it's comparable to extrapolating from the width of a DNA strand to the radius of Earth—and assuming that nothing new happens in between. It's highly questionable that this extrapolation is any good. In any case, if you do it, then the equations eventually just break down; we get the Big Bang scenario, and that's that. It's rather boring, really.

However, because there's no data to constrain this extrapolation back in time, there is nothing to prevent physicists from changing the equations at earlier times and making up exciting stories about what might have happened. That's much more interesting. For example, it is very common for physicists to assume that when densities increase beyond the so-far-tested range, the fundamental forces of nature eventually merge to one in an event called *grand unification*. We have no evidence that something like this ever happened, but a lot of physicists believe it nevertheless. Furthermore, they have come up with hundreds of different ways to change the evolution equations. I cannot possibly go through all of them, but here I'll briefly list the currently most popular ones.

Inflation

According to the theory of **inflation**, the universe was created from quantum fluctuations of a field called the *inflaton*. The word *field* here just means that, unlike a particle, it permeates space and time—it's everywhere. Emergence from quantum fluctuations means that this creation can happen even in vacuum. The universe starts with vacuum, and all of a sudden, there's a bubble with the inflaton field in it, and that bubble keeps expanding. The inflaton field causes the universe to undergo a phase of exponentially fast expansion—the inflation that gives the theory its name. Physicists then postulate that the inflaton field decays into the particles that we still observe today,[a] and from there on, everything continues according to the concordance model.

We have no evidence for the existence of the inflaton field or for the idea that today's particles were produced in its decay. Some physicists have claimed that inflation theory makes predictions that may be falsified by upcoming observations. However, you can always choose the properties of the inflaton field so they match whatever we will observe, which means the hypothesis has no explanatory power. The reason inflation is popular with physicists is that it's believed to simplify the initial conditions, but leaving aside that this claim has been contested, this simplification comes at the cost of complicating the evolution equation.

That the inflaton field gives rise to a universe where previously there was only vacuum is, on occasion, interpreted as creation ex nihilo, "out of nothing," as, for example, in physicist Lawrence Krauss's book *A Universe from Nothing*. A quantum vacuum, however, is not nothing. It is definitely something with very specific mathematical properties. Also, in the common version of inflation theory, space and time existed before the creation of our universe, so it is clearly not creation ex nihilo.

---

[a] This usually includes the hypothetical particles that make up dark matter.

New Forces

Physicists currently count four fundamental forces: gravity, the electromagnetic force, and the strong and weak nuclear forces. All other forces we know of—van der Waals forces, friction, muscle forces, and so on—arise from those four fundamental forces. Physicists call any hypothetical new force a *fifth force*. This name doesn't (yet) refer to any specific force but to a large number of different forces that have been conjectured for different reasons, one of which is to alter the hypothetical conditions in the early universe.

I'll just pick out one for illustration, the force created by a field, the *cuscuton*, that supposedly existed in the early universe. It has since disappeared, but back then it allowed fluctuations to travel faster than the speed of light. The cuscuton is not named after couscous, and not after the marsupial species *cuscus* either, but rather after the plant genus *Cuscuta*. This parasite grows on plants and bushes and looks somewhat like a fuzzy green wig. *Cuscuta* is found almost exclusively in tropical and subtropical regions, which is my excuse for never having heard of it before. The cuscuton field is so named because, like the parasite, the field "grows" on the dynamic law of the concordance model.

The force created by the cuscuton has a similar consequence for the distribution of matter in the universe as the exponential expansion of inflation theory, and it suffers from the same problem—namely, that it is unnecessary to explain any existing observation and provides no simplification over the concordance model.

The cuscuton was first proposed in 2006, and I have to admit it's somewhat of a niche idea. I am mentioning it here because it has been shown that as far as current observations are concerned, the cuscuton can't be distinguished from inflation. This drives home my point that these hypotheses are ambiguous and make a simple story more complicated, the opposite of what scientific theories should do.

Bounces and Cycles

This class of theories has it that the current expansion of our universe was preceded by a contraction phase; they replace the Big Bang with a Big Bounce: that is, a smooth transition from an earlier universe into ours. In some variants of these theories, our universe will eventually end in yet another bounce, part of an infinite cycle. There are various versions of such cycles, depending on just how you change the evolution equation around the Big Bang singularity.

The most popular cyclic models are *conformal cyclic cosmology*, proposed by Roger Penrose, and the *ekpyrotic universe*, originally proposed by Justin Khoury and collaborators. Penrose glues the late phase of the universe to the early phase of the next universe, whereas Khoury and friends imagine that the universe was created in an extradimensional collision of high-dimensional surfaces, which can happen repeatedly. A Big Bounce without a cycle also happens in some approaches that aim to unify gravity with quantum mechanics, like loop quantum cosmology.

The problem with these ideas—you probably guessed it—is that they have no explanatory power. They do not simplify the calculation of any observation; instead, they make things *more* complicated, and it is highly questionable that there is any observation that can ever be uniquely attributed to one of them.

The No-Boundary Proposal

The no-boundary proposal avoids the Big Bang singularity by replacing time with space outside the early universe. I say *outside* because it makes little sense to use *before* if there was no time. Imagine a paper with a circle drawn on it. The circle is our universe as we know it. It has space and time. The area outside the circle has no time. It is not before anything, but next to everything. In the no-boundary proposal, our universe is embedded into space just like that.

This idea was originally proposed by Stephen Hawking and Jim Hartle, but a similar disappearance of time has appeared more recently in some versions of loop quantum cosmology. Yes, that's the same approach to quantizing space-time that, according to other people, might give rise to a bounce. This ambiguity doesn't appear merely because the math is difficult, though it is, but also because there are different ways to turn ideas into math but no data to tell us which is the right way.

Like the other theories for the early universe, this one, too, works by replacing the evolution equation with a different one. The no-boundary proposal suffers from the same problem as all other theories for the early universe: it is unnecessary to explain any observation, it does not result in any simplification, and its predictions are ambiguous.

Geometrogenesis

The idea of *geometrogenesis* ("birth of geometry") is that space was created along with the universe. In such an approach, scientists typically describe the prenatal phase of the universe as some kind of network that has too many connections to lend itself to a meaningful geometric interpretation. This network then changes with time or with temperature and eventually takes on a regular, geometric shape that approximates the space of Einstein's theory.

Geometrogenesis is inspired by the observation that every surface we think of as smooth and continuous—like paper or plastic—upon close inspection is actually made of smaller things and has holes in it. The problem with geometrogenesis is, once again, that it isn't actually necessary to describe anything we observe. It is filling a story into a gap in our knowledge because scientists are unwilling to accept that the answer is "We don't know."

o o o

Let me be clear that I am not saying these models make no predictions. Physicists have all read their Karl Popper, and they usually try to predict something. The problem is that the models are malleable, and if an observation doesn't fit a prediction, that can easily be remedied by amending the models. If physicists hadn't dropped their philosophy of science course after Popper, they'd see the problem with this method. But they don't, which is why we now have hundreds of stories about the beginning of our universe, none of which is actually necessary to explain anything we have observed.

My intention here is not to trash cosmology. OK, maybe a little bit. But we should keep in mind that we have learned some truly amazing facts about the universe from research in cosmology. A century ago, we knew neither that there are galaxies besides our own nor that the universe expands, and I certainly do not want to belittle these achievements. Neither do I want to argue that cosmology is finished. The best current model of the universe, the concordance model, will almost certainly not be the last word. It is foreseeable that data will continue to get better for a long time. This will rule out some models—maybe the concordance model among them—and new, better ones will be put forward and become established. These better models will have good chances to extend further back in time than the concordance model.

Nevertheless, cosmological research is limited by two different problems. First, all these hypotheses about the early universe—the ones I've listed and many others you may have heard about—are pure speculation. They're modern creation myths written in the language of mathematics. Not only is there no evidence for them, but also it's hard to conceive of *any* evidence that could settle the debate regarding which one is correct, because they are all so flexible they can plausibly be made to accommodate any data thrown at them.

Second, when it comes to explaining the early universe, physicists are faced with a fundamental problem that might be impossible to overcome. All our current theories rely on simple initial conditions. This isn't optional; it's essential for our mode of explanation to work. If you have to make the initial conditions complicated, even the simplest evolution law will not give your theory explanatory power. If the universe went through an earlier phase that is more difficult to describe than that hot plasma from which galaxies formed, then our entire scientific methodology would stop functioning. Even if this hypothesis were right, we'd have no rationale that would allow us to add a more difficult story before a simple one.

The only way I can think of to overcome this impasse is to eventually develop theories that do not require initial conditions but instead apply to all times at once. There isn't any such theory at the moment, so that, too, is pure speculation.

In the End

If we take our current theories of the universe and extrapolate them into the distant future, the result, in one word, is dark. In about four billion years, our neighboring galaxy Andromeda is projected to collide with the Milky Way. Our own Sun will have spent its nuclear fuel and burned out in about eight billion years, and so will, eventually, all other stars. While matter cools and clumps, with much of it ending up in black holes, the expansion of the universe will happen faster and faster, making it more and more difficult to see the faint glow of other galaxies as they recede from us. Night skies will go black.

But no one will be around to see them anyway. The universe can support life only in the limited, blessed window of time we currently find ourselves in. That's regardless of how flexibly you define *life*, because the supply of useful energy will inevitably run out. Even if we imagine forms of life very different from ours (Freeman Dyson, for example, speculated that life might form in interstellar clouds of gas), they will all ultimately fall victim to the same problem: life requires change, and change requires free energy, and there's a limited supply of it. Another way to say this is that entropy cannot decrease. We will talk more about entropy in chapter 3. For now, let us just have a critical look at how much one should trust these extrapolations into the far future.

Let me begin by noting that we don't know whether the laws of nature will remain the same even tomorrow. In science, it's often an unwritten article of faith that the laws of nature will remain what they are and not suddenly change.

David Hume, in the eighteenth century, called it the *problem of induction*: when we infer the probability of a future event from past observations, we implicitly assume nature is uniform, constant, and reliable in its proceedings. The laws of nature don't suddenly change. If they did, we wouldn't call them laws.

But we may be mistaken in our assumption that nature is uniform. Bertrand Russell, in his 1912 book *The Problems of Philosophy*, compared Hume's argument to a chicken's attempt at inferring the laws of living on a farm. The chicken is fed reliably every morning at 9:00 a.m., until one day the farmer chops off its head. "More refined views as to the uniformity of nature would have been useful to the chicken," Russell mused.

Hume's eighteenth-century problem is still a problem today, and it might be an unsolvable problem. The uniformity of nature itself is certainly an expectation based on our past observations, but we can't use an assumption to confirm itself. It's impossible to predict that nothing unpredictable will happen.

In case you were hoping that requiring the laws of nature to be mathematical is a way out: sorry, but that doesn't help. It isn't difficult to come up with mathematical laws that will look indistinguishable from the ones we have confirmed so far but will blast apart the solar system tomorrow. It's not that anything speaks for this, but nothing speaks against it either. A smarter chicken might have been able to infer the farmer's intentions, but it would still not have been able to infer that its inference would work.

What is going on? For 97 percent of all Wikipedia articles, if you click on the first link and repeat this in each subsequent article, you will eventually get to an entry about philosophy. Philosophy is where our knowledge ends, and the scientific method is no exception. Does the scientific method work? Yes. Why does it work? Ultimately, we don't know. And because we don't know why it works, we can't be sure it'll continue to work.

Why then do science at all? Why, indeed, do anything when the universe might fall apart any moment? When I first learned about Hume's problem of induction, as an undergrad, I was stumped. I felt that someone had pulled the carpet of reality out from under me, to reveal a big, gaping void. Why hadn't anyone warned me of this?

But then I thought, "Well, what difference does it make?" The laws of nature will either continue to do what they've been doing so far, or they won't. If they continue, the scientific method will serve us well and will help us decide which course of action best suits our needs. If the laws don't continue, there isn't anything we can do about it, and no course of action will prepare us for it, so why bother thinking about it? I rolled back the carpet. There's still a void under it, but I can live with that. I guess I wasn't meant to be a philosopher.

I have the same reaction to scary stories about the demise of our universe. If we can't do anything about it anyway, it's pointless to fret about it.

Take, for example, the risk that the universe might undergo spontaneous vacuum decay, which means the vacuum might suddenly fall apart into particles that come out of nowhere. If that happens, an enormous amount of energy will be released into what was previously empty space. All matter will be ripped apart instantaneously. We cannot rule out this possibility, because observations merely tell us that the vacuum has not decayed so far. This means we cannot tell a truly stable vacuum from one that is merely very long-lived, or *metastable*, as the physicists say. It's Russell's chicken for vacuum-expectation values rather than food-expectation values.

Stickers that glow in the dark, for example, work with metastable states. The paint used for them contains atoms capable of phosphorescence. If you shine light onto these atoms, they temporarily store it by moving electrons to higher, metastable energy levels. When the electrons decay back to the lower level, the atoms release the energy again in the form of light, hence the glow.

Like one of those phosphorescent atoms, our vacuum might also undergo decay. And because this is a quantum process, it's not as if it starts slowly, so we'd see it coming. It just happens with a certain probability within a certain amount of time, with no advance warning.

Whether or not our vacuum can decay depends on a couple of parameters whose values we don't know exactly. The best current estimates say that, yes, the universe can decay, but its average lifetime is something like 10^{500} years. That's a number so big, it doesn't even have a name. But that's only the *average* lifetime. It means the probability is small that the vacuum will decay much earlier than that. But the vacuum *can* decay earlier; it's just very unlikely.

In my opinion, though, this and similar estimates are meaningless, because they require an extrapolation over more than a dozen orders of magnitude of unknown physics, down to distances of about 10^{-35} meters, whereas the best current experiments reach down to only about 10^{-20} meters.[a] If there is anything we don't yet know of in this range (which we have good reason to think is the case), the estimate is wrong. Hence, the brief summary is that we don't know.

Similar considerations apply to other stories about the end of the universe. We can certainly take the laws of nature that we know and extrapolate them, and that's a fun exercise. But even leaving aside the problem of induction, the further we look ahead, the more uncertain our predictions become. If there are any physical processes that are so slow or rare that we haven't observed them so far, they might become relevant in the distant future.

For example, a lot of physicists have speculated that protons, one of the constituents of atomic nuclei, might be unstable but are just so long-lived that we haven't seen one decaying yet. Maybe so, maybe not. Black hole evaporation, too, happens so slowly that we can't measure it—if it happens at all, for which we have no evidence.

We also don't know what dark energy will do in the distant future. We haven't found evidence that its amount changes, but if it changes really slowly, we won't be able to measure it. Yet even an exceedingly slow change in the amount of dark energy would have a large effect

---

[a]A distance of 10^{-35} meters is the so-called *Planck length*, the scale at which quantum gravity is expected to become important, and 10^{-20} meters is approximately the distance probed by the currently largest particle collider in the world, the Large Hadron Collider at CERN.

on the expansion rate. Indeed, when the universe was five billion years younger—a time when our planet hadn't been born but life was already possible on other planets—we probably wouldn't have been able to measure dark energy at all. Back then, the influence of dark energy was much smaller, not large enough to cause the universe's expansion to accelerate.

Lawrence Krauss has joked that he makes predictions only trillions of years into the future, because no one will be around to check if he's correct. It seems to me that the more reliable but less funny prediction is that Krauss won't be around in case it turns out to be wrong that no one will be around. In any case, you shouldn't trust physicists' predictions for the end of the universe. You might as well ask a fruit fly for a weather forecast.

>> THE BRIEF ANSWER

We improve scientific theories by simplification. When it comes to the early universe, there may be a limit to how much we can possibly simplify our explanations. It could therefore be that we will never be able to tell which one of many possible theories for how the universe began is correct. This is certainly presently the case for theories about the beginning of the universe. For possible ways the universe could end, the problem is that we don't know anything about processes that are so rare or slow we wouldn't yet have been able to observe them. So don't take these stories too seriously, but feel free to believe them if you want.

IS MATH ALL THERE IS?

An Interview with Tim Palmer

In autumn 2018, I got a surprise invitation from the Royal Society in London. They asked me to attend a dinner conversation about artificial intelligence. When I looked up the sender, the then-acting president of the society, he turned out to be a Nobel Prize winner. Because my knowledge about artificial intelligence barely extends beyond its being commonly abbreviated AI, I assumed the invitation was a mistake. I didn't respond.

A few weeks passed. Then came a polite reminder to please RSVP. I wrote back to say they had gotten the wrong person. I was assured they indeed wanted me to come. No, really. And I thought, "Well, free trip to London, dinner included." Would you have said no?

This is how I found myself one February evening in the building of the Royal Society, at a big oval of a table, feeling misplaced among people loaded with titles and awards. As I awkwardly sat down, the British gentleman next to me introduced himself as a climate scientist, attending because his group at University of Oxford uses artificial intelligence to study clouds. His name: Tim Palmer, one of the recipients of the

2007 Nobel Peace Prize for his work in the Intergovernmental Panel on Climate Change.

I did not recall it at the time, but the same Tim Palmer had sent me an email a year earlier, about which I joked to my husband that now even climate scientists have ideas for how to revolutionize quantum mechanics. Indeed, after the dinner, Tim tried to initiate a conversation with me about free will in quantum mechanics, of all things. I excused myself and left him standing in a cold, dark London street.

But Tim Palmer, it turned out, is not one to give up easily. He kept sending me cheerful updates about his newest attempts to fix quantum mechanics. I did my best to ignore him, and probably would have succeeded if I had not, a few months later, looked for a climate scientist to interview for an article I was writing.

A year later, we'd written a paper, published a popular-science article along with it, and recorded a song together. Tim and I, it turned out, had independently arrived at similar conclusions about the lack of progress in the foundations of physics. We both pointed the finger at physicists' overreliance on **reductionism**, the idea that we gain deeper insights into nature by looking at shorter and shorter distances. Because the questions about how much we really know and how much we can possibly know are a running theme of this book, I went to interview him again, this time in his office at the University of Oxford.

o o o

Enter Tim's office and a cardboard Einstein greets you at the door, leaning on a whiteboard with scribbles of the Navier-Stokes equation, the math that describes turbulence in the atmosphere. That's Tim's passion in a nutshell—space-time geometry and chaos theory combined. Behind his desk, a European flag mourns the UK's departure from the European Union.

I hesitate for a moment with my first question. Scientists often give me funny looks for it. Still, I think it provides relevant context, so I begin by asking if he is religious.

"No. No, I'm not," Tim says. He shakes his head and his Einstein hair wiggles. Then he adds, "Well, I'm not religious, but I get slightly resistant to people who are adamant they can prove that God doesn't exist." He complains for a bit about scientists like Richard Dawkins who portray all religious people as stupid, ignorant, or both. I realize that there are quite a few of these scientists.

"The reason this bothers me a bit," Tim continues, "is that I know there are a lot of creationists in the US who've been very vocal and all that, but you have to remember that a lot of traditional Muslim families also have this creationist belief. And I was brought up a Catholic, so I am aware there is an element in that that is attacking your culture. It bothers me a bit that this sort of attitude toward creationism could be alienating young people from those cultures that might otherwise might have been open to a career in science.

"So I tried to think, 'Could one envision a situation in which such a belief, that God created the universe six thousand years ago, wasn't stupid and wasn't completely against all the things that we understand about science?'"

I agree with Tim that scientists sometimes overstep the boundaries of their discipline. Of course some religious beliefs have turned out to be just incompatible with evidence. Humans, for example, did not inhabit Earth together with dinosaurs, and having sex in public doesn't increase the banana harvest. But science has limits, and rather than proclaiming that teaching religion is "child abuse"—as Lawrence Krauss has—I think scientists should acknowledge that science is compatible with many traditional sacred beliefs.

Tim goes on to make his case: "The standard argument is that the idea that the universe was created six thousand years ago is stupid because we know that the age of the Earth is billions of years, and the age of the stars is longer than that, and all kinds of lines of evidence make it completely obvious that the universe is much older than six thousand years.

"But then I started thinking, 'What do we mean by this word *creation* anyway?' Let us look, for example, at the creation of atoms. What are atoms? Well, all that science can say at the moment is that we can describe atoms with equations. We have laws that are mathematical, and whatever you want to know about an atom, the equations will tell you what it does. But the mathematics will not tell you what an atom *is*. Is an atom *just* mathematics? Is mathematics all that is? Or is there something, a substance or something, that makes stuff real and is not part of the modern-day scientific canon?

"And the answer is, no one knows. Hawking in his book [*A Brief History of Time: From the Big Bang to Black Holes*] famously asked the question, 'What breathes fire into the equations to make the universe?' Perhaps there is something to the universe around us that isn't just mathematics.

"I am not trying to advocate this," Tim cautions, "but you could say God created the universe as a piece of mathematics. And that mathematics will describe how clumps of dust aggregate and get hot enough that nuclear fusion starts to make energy and elements and so on. All of this is just mathematics. And then, six thousand years ago, God got fed up and said, 'This is a bit boring. I'm gonna make some real stuff now,' and waved his wand and at that point real stuff appeared.

"I was wondering, 'How would science deal with this? What is there in science that would distinguish the pre-creation and post-creation era?' Nothing. Chemistry is underpinned by physics, and that is underpinned by mathematics. So there is nothing in science that would say anything about this moment of creation.

"So I thought if someone was brought up with this belief that creation happened some thousand years ago, here is an easy way out. Six thousand years ago, God created the universe, and before it was all just mathematical equations. And this is not unscientific. It does not go against anything in our current scientific lexicon. I like to use the word *ascientific*. Science has nothing to say about it—at least, science in its current state. There are things we are really profoundly ignorant about. And this is one of them. Is mathematics just a tool for describing the world, or is it the world? We can argue about it, but there is nothing scientific we can say about it."

I ask Tim for other examples where we fill in gaps in our scientific knowledge with belief, and he names the Big Bang. "It's a situation where we have no means of distinguishing between a God-type solution and a scientific one. Unless we find a better theory, maybe one in which there was an earlier eon."

He is, of course, thinking about his own theory, which does away with the division into initial law and differential equation that physicists currently use. Instead, Tim argues, we should describe the universe and everything in it by using the arrangement of matter in the universe, at all times, in its entirety. The geometry of this arrangement could bring new insights into which configurations of particles are even possible, and how likely they are to ever repeat.

This idea led Tim to a theory in which the universe has no beginning and no end. The mathematics for his timeless structure of natural law is a fractal, a pattern of infinite variety in which the large scales resemble the smallest but never exactly repeat. On this fractal, our universe goes through eons that resemble one another but never quite repeat. It has done this for an eternity and will continue doing so forever.

"I didn't do this to get rid of God," Tim says. "It's just how it works out. That's the way physics works. You do the math and find what you find."

"So you don't have a Big Bang, but you have a cycle?"

"Well," he says, "the word *cycle* has this connotation that it repeats, and I wouldn't buy into this. In some sense it cycles: it goes from a Big Bang to a Big Crunch to a Big Bang to a Big Crunch, and so on. But the way I think about it is as a path in a state space, which means a space where each point is a configuration of the universe, so it's a very high-dimensional space. And you plot the path of this multi-eon universe in this state space, and the theory tells you that the path is contained within a finite region of state space and it is a fractal. This is what you would expect if the universe as a whole is a chaotic dynamical system. This means there could be a universe in the past or the future which very closely resembles the current one at the current time. I often think about this: if you are agonizing over a decision you made and you are kicking yourself, 'Why did I do that?,' then don't worry, because there will come an eon when you'll be faced with the same situation and you will make the right decision."

"And there will come an eon when you make an even worse decision," I quip.

He nods without the hint of a smile. "You may make an even worse decision. And the other thing that occurs to me is, if you lose a partner, you may not lose that partner forever. They may come back in a future eon."

I know it sounds crazy. But it's compatible with all we currently know.

>> *THE BRIEF ANSWER*

We use mathematics to describe our observations, but we don't know why some math describes reality whereas other math doesn't. One can therefore attribute a moment of creation specifically to the math that describes what we observe, a moment at which the math becomes real. Such a creation event is by construction not observable—otherwise it would have been described by the math already—and is therefore compatible with science.

Chapter 3

WHY DOESN'T ANYONE EVER GET YOUNGER?

The Last Question

In Isaac Asimov's 1956 short story "The Last Question," a slightly drunk man by the name of Alexander Adell gets seriously worried about the energy supply of the universe. He reasons that, while energy itself is conserved, the useful fraction of energy will inevitably run out. Physicists call this useful energy, which can bring about change, *free energy*. Free energy is the counterweight of entropy. As entropy increases, free energy decreases, and change becomes impossible.

In Asimov's story, tipsy Adell hopes to overcome the second law of thermodynamics, which has it that entropy cannot decrease. He approaches a powerful automatic computer, called Multivac, and asks: "How can the net amount of entropy of the universe be massively decreased?" After a pause, Multivac responds: "INSUFFICIENT DATA FOR MEANINGFUL ANSWER."

Adell's worry—the second law of thermodynamics—is familiar to all of us, even if we don't always recognize it for what it is. It's one of the first lessons we learn as infants: things break, and some things that break can't be fixed. It isn't just Mommy's favorite mug that ultimately suffers this fate. Eventually, everything will be broken and unfixable: your car, you, the entire universe.

It seems our experience that things irreversibly break is at odds with what we discussed in the previous chapter, that the fundamental laws of nature are time-reversible. And in this case we can't just chalk this mismatch up to our fallible human senses, because we observe irreversibility in many systems much simpler than brains.

Stars, for example, form from hydrogen clouds, fuse hydrogen to heavier atomic nuclei, and emit the resulting energy in the form of particles (mostly photons and neutrinos). When a star has nothing left to fuse, it dims or, in some cases, blasts apart into a supernova. But we have never seen the reverse. We have never observed a dim star that took in photons and neutrinos and then split heavy nuclei into hydrogen before spreading out to become a hydrogen cloud. The same goes for countless other processes in nature: Coal burns. Iron rusts. Uranium decays. But we never see the reverse processes.

Superficially, this looks like a contradiction. How can time-reversible laws possibly give rise to the evident time-irreversibility we observe? To understand how this can be, it helps to sharpen the problem. All the processes I just described are time-reversible in the sense that we can mathematically run the evolution law backward in time and recover the initial state. That is to say, the problem is not that we cannot run the movie backward; the problem is that when we run the movie backward, we immediately see that something isn't right: shards of glass jump up and fill a window frame, car tires pick up rubber streaks from the street, water drops lift from an umbrella and rise into the sky. Math may allow that, but it clearly isn't what we observe.

This mismatch between our theoretical and intuitive expectations comes from forgetting about the second ingredient we need to explain observations. Besides the evolution law, we need an initial condition. And not all initial conditions are created equal.

Suppose you want to prepare the batter to bake a cake. You put flour into a bowl, add sugar, a pinch of salt, and maybe some vanilla extract. Then you put butter on top, break a few eggs, and pour in some milk. You begin mixing the ingredients, and they quickly turn into a smooth, featureless substance. Once that has happened, the batter won't change anymore. If you keep on mixing, you will still move molecules from one side of the bowl to the other, but on average the batter remains the same. Everything is as mixed up as it can be, and that's it. Basically, our universe will end up like this too: as mixed up as it can be, with no more change, on average.

In physics, we call a state that doesn't change on average—like the fully mixed batter—an *equilibrium state*. Equilibrium states have reached maximum entropy; they have no free energy left. Why does the batter come into equilibrium? Because it's likely to happen. If you turn on the mixer, it's likely to mash the eggs into the flour but very unlikely to separate the two. This would also happen without the mixer, because the molecules in the ingredients don't sit entirely still, but it would take much longer.[a] The mixer acts like a fast-forward button.

It's the same for the other examples: they are likely to happen only in one direction of time. When pieces of a broken window fall to the ground, their momentum disperses in tiny ripples in the ground and shock waves in the air, but it is incredibly unlikely that ripples in the ground and the air would ever synchronize in just the right way to catapult the broken glass back into the right position. Sure, it's possible mathematically, but in practice it's so unlikely, we never see it happening.

---

[a] In theory. In practice, the eggs would rot long before that, so please don't try it at home.

The equilibrium state is the state you are likely to reach, and the state you are likely to reach is the state of highest entropy—that's just how entropy is defined. The second law of thermodynamics, hence, is almost tautological. It merely says that a system is most likely to do the most likely thing, which is to increase its entropy. It is only *almost* tautological because we can calculate the relation between entropy and other measurable quantities (say, pressure or density), making relaxation to equilibrium quantifiable and predictive.

It sounds rather unremarkable that likely things are likely to happen. Pots break irreversibly because they're unlikely to unbreak. Duh. That's not exactly a deep revelation. But if you pursue this thought further, it reveals a big problem. A system can evolve toward a more likely state only if the earlier state was less likely. In other words, you have to start from a state that's not in equilibrium to begin with. The only reason you can prepare a batter is that you have eggs and butter and flour, and those are not already in equilibrium with one another. The only reasons you can operate a mixer is that you are not in equilibrium with the air in your room[a] and our Sun is not in equilibrium with interstellar space. The entropy in all these systems isn't remotely as large as it could be. In other words, the universe isn't in equilibrium.

Why is that? We don't know, but we have a name for it: the *past-hypothesis*. The past-hypothesis says that the universe started out in a state of low entropy—a state that was very unlikely—and that entropy has gone up ever since. It will continue to increase until the universe has reached the most likely state, in which nothing more will change, on average.

---

[a]Assuming the air temperature is above or below body temperature, because then you'd be dead if you were in equilibrium with it. If it happens to be at body temperature, I applaud your endurance.

For now, entropy can remain small in some parts of the universe—like in your fridge or, indeed, on our planet as a whole—provided these low-entropy parts are fed with free energy from elsewhere. Our planet currently gets most of its free energy from the Sun, some of it from the decay of radioactive materials, and a little from plain old gravity. We exploit this free energy to bring about change: we learn, we grow, we explore, we build and repair. Maybe at some point in the future we will succeed with creating energy from nuclear fusion ourselves, which will expand our capacity to bring about change. That way, if we smartly use the available free energy, we might manage to keep entropy low and our civilization alive for some billion years. But free energy will run out eventually.

This is why the universe has a direction forward in time, the *arrow of time*—it's the direction of entropy increase; it points one way and not the other. This entropy increase is not a property of the evolution laws. The evolution laws are time-reversible. It's just that in one direction the evolution law brings us from an unlikely to a likely state, and that transition is likely to happen. In the other direction, the law goes from a likely to an unlikely state—and that (almost) never happens.

So why doesn't anyone ever get younger? The biological processes involved in aging and exactly what causes them are still the subject of research, but loosely speaking, we age because our bodies accumulate errors that are likely to happen but unlikely to spontaneously reverse. Cell-repair mechanisms can't correct these errors indefinitely and with perfect fidelity. Thus, slowly, bit by bit, our organs function a little less efficiently, our skin becomes a little less elastic, our wounds heal a little more slowly. We might develop a chronic illness, dementia, or cancer. And eventually something breaks that can't be fixed. A vital organ gives up, a virus beats our weakened immune system, or a blood clot interrupts oxygen supply to the brain. You can find many different diagnoses in death certificates, but they're just details. What really kills us is entropy increase.

○　○　○

So far, I have just summarized the currently most widely accepted explanation for the arrow of time, which is that it's a consequence of entropy increase and the past-hypothesis. Now let us talk about how much of this we actually know and how much is speculation.

The past-hypothesis—that the initial state of the universe had low entropy—is a necessary assumption for our theories to describe what we observe. It's a good explanation so far as it goes, but we don't currently have a better explanation than just postulating it. The question why an initial state was what it was just isn't answerable with the theories we currently have. The initial state must have been *something*, but we can't explain the initial state itself; we can only examine whether a specific initial state has explanatory power and gives rise to predictions that agree with observations. The past-hypothesis is a good hypothesis in the sense that it explains what we see. However, to explain the initial state by something else than a yet earlier initial state, we would need a different type of theory.

Of course, physicists have put forward such different theories. In Roger Penrose's conformal cyclic cosmology, for example, the entropy of the universe is actually destroyed at the end of each eon, so the next eon starts afresh in a low-entropy state. This does indeed explain the past-hypothesis. The price to pay is that information is also destroyed for good. Sean Carroll thinks that new, low-entropy universes are created out of a larger multiverse, a process that can continue to happen indefinitely. And Julian Barbour posits that the universe started from a "Janus point" at which the direction of time changes, so actually there are two universes starting from the same moment in time. He argues that entropy isn't the right quantity to consider and that we'd be better off thinking about complexity instead.

You probably know what I am going to tell you next: these ideas are all well and fine, but they're not backed up by evidence. Feel free to believe them—I don't think any evidence speaks against them either—but keep in mind that at this point they're just speculation.

I do have a lot of sympathy for Julian Barbour's argument, however. Not so much because, according to Barbour, time changes direction (which I have no strong opinion about), but because I also don't think entropy is of much use when describing the universe as a whole. To see why, I first have to tell you about the math I swept under the rug with the vague phrase "on average."

o o o

Entropy is formally a statement about the possible configurations of a system that leave some macroscopic properties unchanged. For the batter, for example, you can ask how many ways there are to place the molecules (of sugar, flour, eggs, and so on) in a bowl so you get a smooth batter. Each such specific arrangement of the molecules is called a *microstate* of the system. A microstate is the full information about the configuration: for example, the position and velocity of all those single molecules.

The smooth batter, on the other hand, is what we call a *macrostate*. It's what I referred to earlier loosely as the average that doesn't change. A macrostate can come about by many different microstates that are similar in some specific sense. In the batter, for example, the microstates are all similar in that the ingredients are approximately equally distributed. We choose this macrostate because we can't distinguish one approximately equal distribution of molecules in the batter from any other. For us, they're all pretty much the same.

The initial state, in which the eggs are next to the butter and the sugar is on top of the flour, is also a macrostate, but it's very different from the batter—you can clearly distinguish the state before mixing from that after mixing. To get the state before mixing, you'd have to put the molecules in the right regions: egg molecules in the egg region, butter molecules in the butter region, and so on. The molecules are ordered in this initial state, whereas after mixing they aren't ordered anymore. This is why entropy increase is also often described as the destruction of order.

The mathematical definition of *entropy* is a number assigned to a macrostate: the number of microstates that can give rise to it. A macrostate that you can get from many microstates is likely, hence the entropy of such a macrostate is high. A macrostate that can come about from only comparably few microstates, on the other hand, is unlikely and has low entropy. The mixed batter, in which the molecules are randomly distributed, has many more microstates than the initial, unmixed batter. Thus, the mixed batter has high entropy; the unmixed one, low.

To give you a visual idea of why this is, suppose we have only two ingredients, and we don't have 10^{25} or so molecules but only 36, half of them flour, the other half sugar. I have drawn these in a grid, and marked each flour molecule with a gray square and each sugar molecule with a white square (figure 4). Initially the two substances are cleanly separated: flour at the bottom, sugar on top (figure 4a). Now let us simulate the mixer by randomly exchanging the positions of two adjacent squares, either horizontally or vertically. I have drawn a first step so you see how this works (figure 4b).

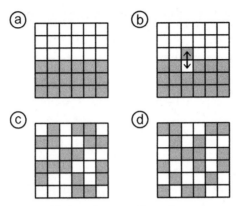

Figure 4: Simple mixing model. Gray squares are flour; white squares are sugar. Mixing switches two neighboring squares randomly.

If we continue to randomly swap neighbors, the molecules will eventually be randomly distributed (figure 4c). What happens is not that the molecules stay in the same place, but they remain equally mixed up. After some more mixing they might look like they do figure 4d. That is, a large number of random swaps gives the same average distribution as if you'd randomly thrown the molecules into the bowl. So instead of having to think about exactly what the mixer does, we can instead just look at the difference between the initial and final distribution.

Let us then define a macrostate of a smooth batter as one in which the sugar and flour squares are approximately equally distributed on top and bottom, say 8 to 10 sugar molecules in the top half (as in figures 4c and 4d). The relevant point is now that there are many more microstates that belong to this macrostate than there are for the initial, cleanly separated state. Indeed, if you don't distinguish molecules of the same type, there's only the initial microstate I drew on the top left, whereas there are many final microstates that are approximately evenly distributed.

This is why entropy is larger in the approximately even distribution, and also why the two substances are unlikely to spontaneously unmix again—it'd require a very specific sequence of random swaps. The sequence required for unmixing becomes less likely the more molecules you mix. Soon it becomes so unlikely that the probability for it to happen in a billion years is ridiculously tiny—you never see it happening.

○ ○ ○

Now that you know how *entropy* is formally defined, let us have a closer look at this definition: entropy counts the number of microstates that can give rise to a certain macrostate. Notice the word *can*. The state of a system is always in only one microstate. The statement that it "can" be in any other state is counterfactual—it refers to states that do not exist in reality; they exist only mathematically. We consider them just because we do not know exactly what the true state of the system is.

Entropy thus is really a measure of our ignorance, not a measure for the actual state of the system. It quantifies which differences between microstates we think aren't interesting. We don't think the specific distribution of the molecules in the batter is interesting, so we lump them together in one macrostate and declare that "high entropy."

This kind of reasoning makes a lot of sense if you want to calculate how quickly a system evolves into a particular macrostate. It therefore works well for all the purposes that the notion of entropy was invented for: steam engines, cooling cycles, batteries, atmospheric circulation, chemical reactions, and so on. We know empirically that it describes our observations of these systems just fine.

This reasoning is, however, inadequate if we want to understand what happens with the universe as a whole, and that's for three reasons. First, and in my mind most important, it's inadequate because our notion of a macrostate implicitly defines already what we mean by *change*. A state that has reached maximum entropy, according to

our definition of a macrostate, still changes (you are still moving batter from one side of the bowl to the other even when it already looks
smooth). It's just that, according to our current theories, this change
is irrelevant. We don't know, however, whether this will remain so
with theories we may develop in the future.

I have illustrated what I mean in figure 5. You can think of these as
two possible microstates at the end of the universe, ten lonely particles
that are randomly distributed in empty space. If the first microstate
(left) changed to the second (right), you wouldn't call that much of a
change. You'd average over it and lump them both together in the same
macrostate.

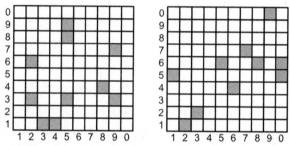

Figure 5: Example for states that superficially look random and very similar but are really
highly ordered and very different from each other.

But now have a closer look at the locations of these particles on the
grid. In the example on the left, they are located at (3,1), (4,1), (5,9),
(2,6), (5,3), (5,8), (9,7), (9,3), (2,3), and (8,4). In the example on the
right, they are at (0,5), (7,7), (2,1), (5,6), (6,4), (9,0), (1,5), (3,2),
(8,6), and (0,6). The super-nerds among you will immediately have
recognized these sequences as the first twenty digits of π and of γ (the
Euler-Mascheroni constant). The distribution of these particles might
look similar to our eyes, but a being with the ability to grasp the sequence of the distribution could clearly distinguish them; they were
created by two entirely different algorithms.

Of course, this example is an ad hoc construction and not applicable to our actual theories, but it illustrates a general point. When we lump together "similar" states in a macrostate, we need a notion of "similarity." We derive this notion from current theories that are based on what we ourselves think of as similar. But change the notion of similarity and you change the notion of entropy. To borrow the terms coined by David Bohm, the *explicate order*, which our current theories quantify, might one day reveal an *implicate order* that we have missed so far.

To me, that's the major reason the second law of thermodynamics shouldn't be trusted for conclusions about the fate of the universe. Our notion of entropy is based on how we currently perceive the universe; I don't think it's fundamentally correct.

There are two more reasons to be skeptical of arguments about the entropy of the universe. One is that counting microstates and comparing their numbers becomes tricky if a theory has infinitely many microstates, and that's the case for all continuous-field theories. It is possible to define entropy in those cases, but whether it's still a meaningful quantity is questionable. It's generally a bad idea to compare infinity to infinity, because the outcome depends on just how you define the comparison, so any conclusion you draw from such an exercise becomes physically ambiguous.

Finally, we don't actually know how to define entropy for gravity or for space-time, but this entropy plays a most important role in the evolution of the universe. You might have noticed that, according to our current theories, matter in the universe starts out as an almost evenly distributed plasma. That plasma must have had low entropy according to the past-hypothesis. But I told you earlier that the smooth batter had high entropy. How does this fit together?

It fits together if you take into account the fact that gravity makes the almost even, high-density plasma in the early universe extremely unlikely. Gravity wants to clump things, but for some reason they weren't very clumped when the universe was young. That's why the initial state had low entropy. Once it evolves forward in time, sure enough, the plasma begins to clump, forming stars and galaxies— because that's likely to happen. This doesn't happen in the batter, because the gravitational force isn't strong enough for such a small amount of matter at comparably low density. It's because of the different role of gravity that the batter and the early universe are two very different cases, and why the one has high entropy, the other one low entropy.

However, to make this case quantitative, we'd have to understand how to assign entropy to gravity. While physicists have made some attempts at doing that, we still don't really know how to do it, because we don't know how to quantize gravity.

For these reasons, I personally find the second law of thermodynamics highly suspect and don't think conclusions drawn from it today will remain valid when we understand better how gravity and quantum mechanics work.

o o o

In Asimov's short story, the universe gradually cools and darkens. The last stars burn out. Life as we know it ceases to exist and is superseded by cosmic consciousnesses, disembodied minds that span galaxies and drift freely through space. Cosmic AC, the last and greatest version of the Multivac series, is again tasked with answering the question how to decrease entropy. Yet again, it stoically replies, "THERE IS AS YET INSUFFICIENT DATA FOR A MEANINGFUL ANSWER."

Eventually, the last remaining conscious beings fuse with the AC, which now resides "in hyperspace" and is "made of something that [is] neither matter nor energy." Finally, it finishes its computation.

The consciousness of AC encompassed all of what had once been a Universe and brooded over what was now Chaos. Step by step, it must be done. And AC said, "LET THERE BE LIGHT!" And there was light.

The Problem of the Now

Einstein's greatest blunder wasn't the cosmological constant, and neither was it his conviction that God doesn't play dice. No, his greatest blunder was speaking to a philosopher named Rudolf Carnap about the Now, with a capital n.

"The problem of the Now," Carnap wrote in 1963, "worried Einstein seriously. He explained that the experience of the Now means something special for man, something different from the past and the future, but that this important difference does not and cannot occur within physics."

I call it Einstein's greatest blunder because, unlike the cosmological constant and his misgivings about indeterminism, this alleged problem of the Now still confuses philosophers, and some physicists too.

The problem is often presented like this. Most of us experience a present moment, which is a special moment in time, unlike the past and unlike the future. But if you write down the equations governing the motion of, say, some particle through space, then this particle is described, mathematically, by a function for which no moment is special. In the simplest case, the function is a curve in space-time, which just means the object changes its location with time. Which moment, then, is Now?

You could argue rightfully that as long as there's just one particle, nothing is happening, and so it's unsurprising that no indication of change appears in the mathematical description. If, in contrast, the particle could bump into some other particle, or take a sudden turn, then these instances could be identified as events in space-time. That something happens seems a minimum requirement to meaningfully talk of change and make sense of time. Alas, that still doesn't tell you whether these changes happen to the particle Now or at some other time.

Now what?

Some physicists, for example Fay Dowker, have argued that accounting for our experience of the Now requires replacing the current theory of space-time with another one. David Mermin has claimed it means a revision of quantum mechanics is due. And Lee Smolin has boldly declared that mathematics itself is the problem. It is correct, as Smolin argues, that mathematics doesn't objectively describe a present moment, but our experience of a present moment is not objective—it's subjective. And that subjectivity can well be described by mathematics.

Don't get me wrong. It seems likely to me that we will one day have to replace the current theories with better ones. But understanding our perception of the Now alone does not require it. The present theories can account for our experience; we merely have to remember that humans are not elementary particles.

The property that allows us to experience the present moment as unlike any other moment is memory—we have an imperfect memory of events in the past and we do not have a memory of events in the future. Memory requires a system of some complexity, one with multiple states that are clearly distinguishable and stable for extended periods of time. Our brain has the required complexity. But to understand better what is going on, it helps to leave aside consciousness. We can do this because memory is not exclusive to conscious systems. Many systems much simpler than human brains also have memory, so let us have a look at one of those: mica.

Mica is a class of naturally occurring minerals, some of it as old as a billion years. Mica is soft for a mineral, and small particles passing through it—maybe from radioactive decays in surrounding rocks—can leave permanent tracks in it. This makes mica a natural particle detector. Indeed, particle physicists have used old samples of mica to search for traces of rare particles that might have passed through. These studies have remained inconclusive, but they're not the relevant point here. I am merely telling you this because mica, though it arguably has low levels of consciousness, clearly has memory.

Memories in mica don't fade like ours do. But, like us, mica has a memory of the past and not of the future. That means that at any particular moment, mica has information about what has happened but no information regarding what's about to happen. It would be a stretch to say that mica has experience of any kind, but it keeps track of time—it knows about the Now.

From mica we can learn that if we want to describe a system with memory, just looking at the proper time—as we did in the previous chapter—isn't enough. For each moment of proper time, we need to ask, "What times does the system have a memory of?" The fact that this memory abruptly ends at the proper time itself is why each moment is special as it happens.

If that sounds confusing, imagine your perception of time as a collection of photographs in different stages of fading. The moment you call Now is the photograph that's least faded. The more faded a photograph is, the more it is in the past. You don't have photographs of the future. At each moment, the Now is your most vivid, most recent photo, with a long trail of fading snapshots behind it and a blank for the future.

Of course, this is an overly simplistic description of human memory. Our actual memory is much more complicated than this. To begin with, we retain some memories and not others, we have several different types of memory for different purposes, and sometimes we believe we have memories of things that didn't happen. But these neurological subtleties aren't important here. What's important is that the present moment is special because of its prominent position in your memory. And the next moment is special too: in each moment, your perception of that same moment stands out.

This is why our experience of a Now is perfectly compatible with the block universe in which the past, present, and future are all equally real. Each moment subjectively feels special at that very moment, but objectively that's true for every moment.

We can see, then, that the origin of the problem of the Now is not in the physics, and not in the mathematics, but in the failure to distinguish the subjective experience of being inside time from the timeless nature of the mathematics we use to describe it. According to Carnap, Einstein spoke about "the experience of the Now [that] means something special for man." Yes, it means something special for man; it means something special for all systems that store memory. However, this does not mean, and certainly does not necessitate, that there is a present moment that is objectively special in the mathematical description. Objectively, the Now doesn't exist, but subjectively we perceive each moment as special. Einstein should not have worried.

The upshot—please forgive me—is that Einstein was wrong. It is possible to describe the human experience of the present moment with the "timeless" mathematics we now use for physical laws; it isn't even difficult. You don't have to give up the standard interpretation of quantum mechanics for it, or change general relativity, or overhaul mathematics. There is no problem of the Now.

Incidentally, Carnap answered Einstein's worry about the Now quite as I just did. Carnap remembers remarking to Einstein that "all that occurs objectively can be described in science" but that the "peculiarities of man's experiences with respect to time, including his different attitude towards past, present, and future, can be described and (in principle) explained in psychology."

I'd have said it's explained by neurobiology and added that biology is ultimately also based on physics. (If this upsets you, you will especially enjoy the next chapter.) Nonetheless, I agree with Carnap that it's important to distinguish objective mathematical descriptions of a system from the subjective experience of being part of the system.

o o o

So there is no problem of the Now. But the discussion about memory is useful to illustrate the relevance of entropy increase for our perception of an arrow of time. I told you in the previous section why forward in time looks different from backward in time, but not why it's the direction of entropy increase that we perceive as forward. Mica illustrates why.

The reason mica doesn't have a memory of the future is that creating its memory increases entropy. A particle goes through the mineral and kicks a neatly aligned sequence of atoms out of place. The atoms remain displaced because part of the energy that moved them disperses into thermal motion and maybe some sound waves. In that process, entropy increases. The reverse process would require fluctuations in the mineral to build up and emit a particle that heals a track in the mineral perfectly. That would decrease entropy and is, hence, incredibly unlikely to happen. The entire reason that we see a record in the mineral is that this process is unlikely to spontaneously reverse.

Memory formation in the human brain is considerably more difficult than that, but it, too, goes back to low-entropy states that left traces in our brain. Say you have a memory of your graduation day. It is likely that this event was in the past and created by light that hit your retina. It is incredibly unlikely that the event will instead be in the future and somehow suck the memory out of your brain. Such things just don't happen. And the reason they don't happen is that entropy increases in only one direction of time.

In the long run, of course, further entropy increase will wash out any memory.

o o o

In summary, neither our experience of an arrow of time nor that of a present moment requires changing the theories we currently use. Of course, some physicists have nevertheless put forward proposals for different laws that are bona fide time-irreversible, but such modifications are unnecessary to explain currently available observations. For all we know, the block universe is the correct description of nature.

Many people feel uneasy when they first realize that Einstein's theories imply that the past and future are as real as the present, and that the present moment is only subjectively special. Maybe you are one of them. If so, it is worth combating your uneasiness, because the reward is seeing that our existence transcends the passage of time. We always have been, and always will be, children of the universe.

Brains. In Empty Space.

> We will all come back
> At the end of time
> As a brain in a vat, floating around
> And purely mind.
> —Sabine Hossenfelder, "Schrödinger's Cat"

The realization (!) that reality is but a sophisticated construction our mind produces from sensory input, and that our perception of it can therefore change if the input changes, has made its way into pop culture in movies like *The Matrix* (in which the protagonist is raised in a computer simulation only to discover that reality looks rather different) and *Inception* (in which the protagonists struggle to devise ways of telling dream from reality) and *Dark City* (in which memories are adjusted each midnight), though such accounts tend to shy away from suggesting that reality ultimately does not exist. There are places even Hollywood won't go.

It's not a new idea that you may be just an isolated brain in a vat, or in an empty universe, with sensory input that creates the illusion of being a human on planet Earth. The idea that we can't really know anything for sure besides the fact that we ourselves exist is an old philosophy known as *solipsism*. As so often, the first written record of someone contemplating this possibility comes from a Greek philosopher, Gorgias, who lived about 2,500 years ago. But solipsism is more commonly associated with René Descartes, who summed it up with "I think, therefore I am," adding at length that, of all the other things, he could never be quite certain.

You may have hoped that physics gets you out of this conundrum, but it doesn't. It makes it worse. That's because in my elaboration about entropy increase I have omitted an inconvenient detail: entropy actually doesn't always increase. And if it decreases, weird things happen.

Let us look again at our simplified batter-mixing model with the 36 squares. Suppose you have reached a state of high entropy, a *smooth* macrostate with 8 to 10 gray squares in the top half. Thing is, if you keep on randomly swapping neighbors, the state will not forever stay smooth. Every once in a while, just coincidentally, there'll be only 7 sugar molecules in the upper half. Keep on swapping and you'll come across an instance when there are only 6. It's unlikely to remain so for long, and probably you'll soon get back to a smooth state. But if you just stubbornly keep on mixing the squares, eventually you'll have only 5, 4, 3, 2, 1, and even 0 gray squares in the upper half. You will have gone back all the way to the initial state. Entropy, it will seem, has decreased.

This isn't a mistake; it's how entropy works. After you have maximized it and reached an equilibrium state, entropy can coincidentally decrease again. Small out-of-equilibrium fluctuations are likely; bigger ones, less likely. A substantial decrease of entropy in the mixing of a real batter is so unlikely that you'd not have seen it happening yet even if you'd been mixing since the Big Bang. But if you could just mix long enough, eggs would eventually reassemble and butter would form a clump again. This isn't a purely mathematical speculation either—spontaneous decreases in entropy can, and have been, observed in small systems. Tiny beads floating in water, for example, have been observed to occasionally gain energy from the random motion of the water molecules. This temporarily defies the second law of thermodynamics.

Such entropy fluctuations create the following problem. If we take together everything we know about the universe, it looks as though it'll go on expanding for an infinite amount of time. As entropy increases, the universe becomes more and more boring. Eventually, when all the stars have died, all matter has collapsed to black holes, and those black holes have evaporated, it'll contain only thinly distributed radiation and particles that occasionally bump into one another.

But this isn't the end of the story, because infinity is a really long time. In an infinite amount of time, anything that *can* happen *will* eventually happen—no matter how unlikely.

This means that in that boring, high-entropy universe, there'll be regions where entropy spontaneously decreases. Most of them will be small, but one day there'll come a large fluctuation, one in which particles form, say, a sugar molecule—just coincidentally. Wait some more, and you'll get an entire cell. Wait even more, and eventually a fully functional brain will pop out of the high-entropy soup for long enough to think, "Here I am," and then disappear again, washed away by entropy increase. Why does it disappear again? Because that's the most likely thing to happen.

These self-aware, low-entropy fluctuations are *Boltzmann brains*, after Ludwig Boltzmann, who in the late nineteenth century developed the notion of entropy we now use in physics. That was before the advent of quantum mechanics, and Boltzmann was concerned with purely statistical fluctuations in collections of particles. But quantum fluctuations add to the problem. With quantum fluctuations, low-entropy objects (brains!) can appear even out of vacuum—and then they disappear again.

You might think that brains fluctuating into existence is pushing things a little too far. You wouldn't be alone. The physicist Seth Lloyd said about Boltzmann brains, "I believe they fail the Monty Python test: Stop that! That's too silly!" Or, as Lee Smolin once put it to me, "Why brains? Why isn't anyone ever talking about livers fluctuating into existence?" Fair point. But I side with Sean Carroll; I think that Boltzmann brains have something worthwhile to teach us about cosmology.

The issue with Boltzmann brains isn't so much the brains themselves; it's that the possibility of such large fluctuations leads to predictions that disagree with our observations. Remember that the lower the entropy, the less likely the fluctuation. Entropy has to be small enough to explain the observations you have made so far. At the very least, that would be your brain with all the input you have gotten in your life. But then the theory predicts with overwhelming probability that the next thing you will see is that planet Earth disappears and entropy relaxes back to equilibrium. Well, that obviously didn't happen. It didn't happen again just now. Still hasn't happened. Meanwhile, you have thoroughly falsified the prediction.

Certainly it isn't good if a theory leads to predictions that disagree with observations. Something must be wrong, but what? The shortcomings in our understanding of entropy that I named earlier (gravity, continuous fields) still apply, but there is another assumption in the Boltzmann brain argument that's more likely to be the culprit. It's that not all types of evolution laws give rise to all possible fluctuations.

A theory in which any kind of fluctuation will eventually happen is called an *ergodic* theory. The little batter-mixing model we used is ergodic, and the models that Boltzmann and his contemporaries studied are also ergodic. Alas, it is an open question whether the theories we currently use in the foundations of physics are ergodic.

One hundred fifty years ago, physicists were concerned with particles that bump into one another and change direction, and asked questions like "How long does it take until all oxygen atoms collect in one corner of the room?" That's a good question (answer: a very, very long time; don't worry), but to talk about the creation of something as complex as a brain, you need to get particles to stick together. They must form *bound states*, as physicists say. Protons, for example, are bound states of three quarks, held together by the strong nuclear force. Stars are also bound states; they are gravitationally bound. Just bouncing particles off one another isn't enough to create a universe that resembles the one we actually observe. And no one yet knows whether gravity and the strong nuclear force are ergodic, so there is no contradiction in the Boltzmann brain argument.

Indeed, we can instead read the argument backward and conclude that at least one of our fundamental theories can't be ergodic. That's why I think Boltzmann brains are interesting—they tell us something about the properties that the laws of nature must have. But you don't need to worry that you're a lonely brain in empty space. If you were, you'd almost certainly just have disappeared. Or if you haven't already, then you'll disappear now. Or now . . .

o o o

Boltzmann brains are a theoretical device to lead an argument by contradiction (if the laws of nature were ergodic, then your observations would be incredibly unlikely), but you almost certainly aren't one. However, there is, I think, a deeper message in the paper trail that Boltzmann brains left in the scientific literature.

The foundations of physics give us a closer look at reality, but the closer we look at reality, the more slippery it becomes. Our heavy use of mathematics is a major reason. The more the fundamental descriptions of nature have become divorced from our everyday experience, the more we must rely on mathematical rigor. This reliance has consequences. Using math to describe reality means that the same observations can be equivalently explained in many different ways. That's just because there are many sets of mathematical axioms that will give the exact same predictions for all available data. Thus, if you want to assign "reality" to one of your explanations, you won't know which.

For example, in Isaac Newton's day, arguing that the gravitational force is real would have been uncontroversial. It was an enormously useful mathematical tool to calculate anything from the path of a cannonball to the orbit of the moon. But along came Albert Einstein, who taught us that the effect we call *gravity* is caused by the curvature of space-time; it's not a force. Does this mean the gravitational force stopped existing with Einstein?

Figure 6: Rabbit or duck?

That would mean that what is real depends on what humans believe to be real. Most scientists wouldn't want to go there.

Well, you may say, it's not that the gravitational force stopped existing with Einstein. It never existed in the first place. Pre-Einstein scientists were just wrong! Ah, but in that case you can't claim that anything in our current theories is real, for one day these theories might be replaced by better ones. Space? Electrons? Black holes? Electromagnetic radiation? You would not be allowed to call these real. Again, most scientists would balk at such a notion of reality.

Even leaving aside this problem of impending paradigm shifts, it's ambiguous what mathematics you use to describe observations, because in physics we have *dual* theories. Two theories that are dual describe the same observable phenomena in entirely different mathematical form. Dual theories are like the drawing that, depending on what way you look at it, is either a rabbit or a duck (figure 6). Is it really a rabbit or really a duck? Well, really it's just a dark line on a white background that you can interpret one way or the other.

In physics, the most famous example is the *gauge-gravity duality*. It's a mathematical equivalence that links a higher-dimensional gravitational theory (one with curved space-time) to a particle theory in one dimension less without gravity (e.g., in flat space-time). In both theories you have a prescription to calculate measurable quantities (like, say, the conductivity of a metal). These mathematical elements of the theories (of gravity or particles) are different, and the prescriptions to calculate with them are different, but the predictions are exactly the same.

Now, it's somewhat controversial whether the gauge-gravity duality actually describes something we observe in our universe. Lots of string theorists believe it does. I, too, think there's a fair chance it correctly describes certain types of plasma that are dual to particular types of black holes. (Or are they black holes dual to some kind of plasma?) But whether this particular dual theory correctly describes nature is somewhat beside the point here. The mere possibility of dual theories supports the conclusion drawn from the threat of impending paradigm changes: we can't assign "reality" to any particular formulation of a theory. (The various different interpretations of quantum mechanics are another case in point, but please allow me to postpone this discussion to chapter 5.)

It is because of headaches like this that philosophers have put forward a variant of realism called *structural realism*. Structural realism has it that what's real is the mathematical structure of a theory, not any particular formulation of it. It's the rabbit-duck shape of the drawing, if you wish. Einstein's theory of general relativity structurally contains what was previously called the *gravitational force* because we can derive this force in an approximation called the *Newtonian limit*. Just because that limit isn't always a good description for our observations (it breaks down near the speed of light and when space-time is strongly curved) doesn't mean it's not real.

In structural realism, you can call gravitational forces real even though they're only approximations. You are also allowed to call space-time real, even though it might one day be replaced with something more fundamental—a big network, maybe? Because whatever the better theory is, it will have to reproduce the structure we currently use in suitable limits. It all makes sense.

If I were a realist, I'd be a structural realist. But I am not. The reason is that I can't rule out the possibility that I'm a brain in a vat and that all my supposed knowledge about the laws of nature is an elaborate illusion. I may be able to reason myself to conclude it's implausible I'm a fluctuation in an otherwise featureless universe, given all I have learned in my life, but that still doesn't prove there is any universe besides my brain to begin with. Solipsism may be called a philosophy, but it's born out of biological fact. We are alone in our heads, and, at least so far, we have no possibility to directly infer the existence of anything besides our own thoughts.

Even so, while I contend that I can never be entirely sure anything besides myself exists, I also find it a rather useless philosophy to dwell on. Maybe you don't exist, and it's just my illusion that I've written this book, but if I can't tell an illusion from reality, why bother trying? Reality certainly is a good explanation that comes in handy. For all practical purposes, therefore, I'll deal with my observations as if they were real, granting the possibility—in case someone asks—that I am not perfectly sure either this book or its readers actually exist.

>> THE BRIEF ANSWER

We get older because that's the most likely thing to happen. Our current theories are described well by the one-directional nature of time and our perception of Now. Some physicists consider the existing explanations unsatisfactory, and it is certainly worth looking for better explanations, but we have no reason to think this is necessary, or even possible. If you want to believe you're a brain in a vat, that's all right, but I wonder what difference you think it makes.

Chapter 4

ARE YOU JUST A BAG
OF ATOMS?

What Are You?

Some public speakers, I am told, cope with speech anxiety by picturing their audience naked. I don't know about you, but I'd rather not. I prefer picturing them taken apart into chemical elements (figure 7).

The human body is about 60 percent water, so that makes my audience first of all a lot of oxygen and hydrogen. I imagine it floating away with a puff. Then, for each person, I have a big jar of carbon, a major con-

Figure 7: Major atomic constituents of the human body by mass percentage. Not to scale.

stituent of proteins and fats. Carbon alone makes up about 18 percent of the human body, something like thirty pounds of an average adult. Then we have another gas, nitrogen (3 percent), a few smaller jars for calcium (1.5 percent) and phosphorus (1 percent), and tiny doses of potassium, sulfur, sodium, and magnesium. And that's about it. That's what humans are: pretty much indistinguishable collections of chemical elements.

If that doesn't work for you, maybe it helps to ponder the origin of your atoms. The universe didn't start out with chemical elements in place, except for hydrogen, which was created a few minutes after the Big Bang, because making the atomic nuclei for the chemical elements requires substantial pressure. Heavy elements could be generated only once stars began to form from hydrogen clouds under the pull of gravity. In these collapsing clouds, gravitational pressure eventually ignites nuclear fusion, which merges the cores of light nuclei to increasingly heavier ones.

But there comes a time when a star has fused everything it had to fuse. At the end of their lives, most stars dim calmly, but some of them collapse rapidly and subsequently explode, thereby becoming a *supernova*. The supernova explosion blows the star's interior out into the cosmos. Freed from the busy environment of the star, the released atomic nuclei then catch electrons and become proper atoms.

But even a supernova explosion doesn't entirely annihilate a star; it leaves behind a remnant that is either a neutron star or a black hole. Neutron stars are big blobs of nuclear matter, so dense they just barely escape collapsing to a black hole. The heaviest of the elements, such as gold and silver, can form only in a particularly violent environment, such as neutron star mergers. In these mergers, too, heavy nuclei are blown out and distributed throughout galaxies, where they catch electrons and become atoms.

Some of these atoms come together to form small molecules or even microscopic grains—*stardust*. The dust mixes into clouds of hydrogen and helium, which are still around from the Big Bang. And gravity continues its work. If the clouds get too dense, they will collapse again, give birth to new stars, solar systems, planets, and, potentially, life on these planets.

This process it not cyclic, and to our best current knowledge it cannot continue forever. At some point in the far future—estimated to be about a hundred trillion years from now—the universe's remaining nuclear fuel will be gone for good. This is one of the consequences of entropy increase, which we talked about in chapter 3. The universe can host life for only a limited amount of time.

But here we are, made of atoms that either came straight from the Big Bang or were thrown into interstellar space by stars in their final fit of anger. As the meme has it, we are made of stardust, children of the stars, and so on. Personally, I don't care much where my atoms came from, but at this point I have usually forgotten my speech anxiety.

More Is More

Does it take anything more than particles to make a conscious being?

I have found that many people reflexively reject the possibility that human consciousness arises from interactions of the many particles in their brain. They seem wedded to the idea that somehow something must be different about consciousness. And while the scientifically minded among them do not call it a *soul*, it is what they mean. They are looking for the mysterious, the unexplainable, the Extra that would make their existence special. They find it inconceivable that their precious thoughts are "merely" consequences of a lot of particles doing whatever the laws of nature dictate. Certainly, they insist, consciousness must be more than this. In a 2019 survey, 75.8 percent of Americans subscribed to this idea of *dualism*—that the human mind is more than a complicated biological machine. In Singapore, the percentage of dualists was even higher: 88.3 percent.

If you are among the dualist majority, we have to make a deal before we can move on. You put aside your belief that consciousness requires some Extra that physics doesn't account for and hear what I have to say. In return, I promise that if you, at the end of this book, still insist the human brain is exempt from the laws of nature, I'll let you get away with it.

Having said that, as a particle physicist by training, I have to inform you that the available evidence tells us that the whole *is* the sum of the parts, not more and not less. Countless experiments have confirmed for millennia that things are made of smaller things, and if you know what the small things do, then you can tell what the large things do. There is not a single known exception to this rule. There is not even a consistent theory for such an exception.

Just as a country's history is a consequence of the behavior of its citizens and their interactions with the environment, so is the behavior of the citizens a consequence of the properties and interactions of the particles they are made of. Both are hypotheses that have withstood any test they have been subjected to—so far. As a scientist, I therefore accept them. I accept them not as ultimate truths, for they may one day be revised, but as best current knowledge.

A lot of people seem to think it is merely a philosophical stance that the behavior of a composite object (for example, you) is determined by the behavior of its constituents—that is, subatomic particles. They call it reductionism or *materialism* or, sometimes, *physicalism*, as if giving it a name that ends in *-ism* will somehow make it disappear. But reductionism—according to which the behavior of an object can be deduced from ("reduced to," as the philosophers would say) the properties, behavior, and interactions of the object's constituents—is not a philosophy. It's one of the best-established facts about nature.

Nevertheless, I am not a reductionist hard-liner. Our knowledge about the laws of nature is limited, much remains to be understood, and reductionism may fail in subtle ways I will discuss later. However, you have to learn the rules before you can break them.

And in science, our rules are based on facts. Fact is, we have never observed an object composed of many particles whose behavior falsified reductionism, though this could have happened countless times. We have never seen a molecule that didn't have the properties you'd expect, given what we know about the atoms it is made of. We have never encountered a drug that caused effects that its molecular composition would have ruled out. We have never produced a material whose behavior was in conflict with the physics of elementary particles. If you say "holism," I hear "bullshit."

We certainly know of many things that we cannot currently predict, for our mathematical skills and computational tools are limited. The average human brain, for example, contains about a thousand trillion trillion atoms.[a] Even with today's most powerful supercomputers, no one can calculate just how all these atoms interact to create conscious thought. But we also have no reason to think it is not possible. For all we currently know, if we had a big enough computer, nothing would prevent us from simulating a brain atom by atom.

In contrast, assuming that composite systems—brains, society, the universe as a whole—display any kind of behavior that does not derive from the behavior of their constituents is unnecessary. No evidence calls for it. It is as unnecessary as the hypothesis of God. Not wrong, but ascientific.

---

[a]In case you find paraphrasing large numbers as confusing as I do, that's about 10^{27}.

This may come as a shock to some of you. Didn't Philip Anderson—a Nobel Prize winner!—claim the contrary when he coined the catchphrase "More is different"? Indeed, he did. But just because a Nobel Prize winner said it does not mean it is correct.

o o o

Up until about fifty years ago, physicists described a system at different levels of resolution with different mathematical models. They would, for example, use one set of equations for water, then another set for its molecules, and yet another set for atoms and their constituents. These different mathematical models were independent of one another.

By the middle of the twentieth century, however, physicists began to formally connect these different models. I say *formally* because the mathematical derivations can in most cases not be executed yet; the calculations are just too difficult. But physicists now have a well-defined procedure to derive, say, the properties of water from the properties of atoms. This procedure is called *coarse-graining*, and while the mathematics is tough, the idea is conceptually simple.

Consider that you're describing a system at high resolution, meaning you take into account lots of fine structures at short distances. Imagine, for example, a topographic map, one that tells you not only where mountain ridges and valleys are but that goes all the way down to creases in the asphalt and pebbles in meadows. If you plan a hike, there is a lot of detail in this map that you do not need. To create a map that's better suited for your purposes, you could put, say, a hundred-yard grid on the terrain and use average values for each square of the grid. This would mean you'd discard information, but it would be information you wouldn't need.

Coarse-graining in physics is a more complicated version of this averaging; it's a method for discarding information you don't need. In physics, the size of the grid is often referred to as the *cutoff*, and the task is to write down an approximate model that is accurate enough at the resolution given by the cutoff, plus small corrections for the missing details. If you then throw away the small corrections below the cutoff for good, you have what physicists call an **effective model**. This model is not fundamentally correct—because, like your averaged topographic map, it is missing information—but it is good enough at the level of resolution you are interested in.

The best-known examples for effective models are bulk descriptions of gases and fluids in terms of aggregate quantities like temperature, pressure, viscosity, density, and so on. These descriptions average molecular details. There are many other effective models that we use in physics. It is typical of an effective model that the objects and quantities central to it are not the same as those in the underlying theory; they usually do not even make sense in the underlying theory. The conductivity of a metal, for example, is a property of materials that derives from the behavior of electrons. But it makes no sense to speak of the conductivity of an electron. Indeed, the whole concept of a *metal* makes no sense if you are working with a model of subatomic particles. A metal is a certain arrangement of many small particles.

We say that such properties and objects, which play a key role in the effective theory but do not appear in the fundamental theory, are **emergent**.[a] Emergent properties and objects can be derived from or reduced to something else. **Fundamental** is the opposite of emergent. A fundamental property or object cannot be derived from or reduced to anything else. Two other terms I will use in the following is that

---

[a]To be more precise, this case is called *weak* emergence. Philosophers distinguish it from *strong* emergence, which refers to the hypothetical case of a macroscopic system having properties that are not derivable from its constituents and their behavior. We'll talk more about strong emergence in chapter 6.

the more fundamental layers are the *deeper* ones, whereas the emergent ones are *higher* levels.

Pretty much everything we deal with in everyday life is emergent, i.e., a high-level property or object. The color of a material (high level) emerges from its atomic structure (deeper level). The potency of a drug (high level) emerges from its molecular composition (deeper level), and the molecular composition further emerges from the molecules' atomic composition (even deeper). The motion of a cell emerges from the arrangements and interactions of its molecules. The function of an organ emerges from that of its cells, and so on.

As the example of the coarse-grained topographic map illustrates, in the process of deriving emergent properties, we discard details that reside at short distances. This is why going from one level to the next higher one in the theory tower is a one-way street. You can derive the laws of hydrodynamics (which describe the motions of fluids) from the theory of atoms. You cannot, however, derive the atomic theory from hydrodynamics. That's because in the derivation of the effective model you throw away information for good. This usually happens in the mathematics by taking some parameter to infinity or, equivalently, by discarding small corrections. In fact, that this theory tower is not a two-way street is why we cannot just deduce more fundamental laws from the laws we have. If we could, they wouldn't be more fundamental! (So how do physicists discover more fundamental laws, then? We'll talk about this with David Deutsch in the next interview.)

In most cases, we currently cannot perform the mathematical calculations that would be necessary for coarse-graining. For example, no one can at present derive the properties of a cell from those of its atoms. Indeed, even predicting the properties of molecules is difficult, as the protein-folding problem illustrates. The math is just too difficult.

But it doesn't matter for our purposes whether or not we can actually perform the calculation that connects the deep level with the high level. We are interested here only in what we can learn from the structure of natural laws. Therefore, what matters is merely that, according to well-established theories, the deepest level determines what happens at the higher levels. If someone now claims that this isn't so, they must at the very least explain how this can be. How can it be that a theory for, say, a metal does not follow from the theory for the collection of the metal's constituents? If you want to push this idea, that's the challenge you have to meet.

Emergent theories aren't of any less importance than fundamental ones. Indeed, they tend to be *more* useful exactly *because* they ignore irrelevant details. Emergent theories are in most cases the better explanations at their level of accuracy. But the only fundamental theories we currently know of—the currently deepest level—are the **standard model of particle physics** and Einstein's general relativity, which describes gravitation.

I will in the following refer to the areas of physics that study the fundamental laws as the *foundations of physics*. Everything else emerges from those fundamental laws, roughly in this order: atomic physics, chemistry, materials science, biology, psychology, sociology. Most physicists, myself included, don't think the currently fundamental theories will remain fundamental. More likely, what's currently fundamental will turn out to be emergent from yet another, deeper level.[a]

In hindsight, it might seem patently obvious that scientific disciplines are tied together in this way. But this was not how scientists thought about nature for most of the previous century. Indeed, outside the foundations of physics, you still find many who vehemently argue that all scientific disciplines are equally fundamental.

---

[a]The search for this deeper level was the subject of my previous book, *Lost in Math: How Beauty Leads Physics Astray*, and I will not cover it here in detail.

Now, to some extent this is quibbling over words. I use the term *fundamental* to mean "cannot be derived from another theory." Scientists in other disciplines sometimes think less fundamental implies less important, and then they're insulted. But physicists aren't trying to belittle other scientists by pointing out that everything is made of particles; it's just how it is.

I said I'd be honest with you, so I should add that some physicists still don't believe that natural laws are indeed reductionist. I don't have much to say about this except that I've laid out the evidence, and you can evaluate it yourself. The hypothesis that nature is reductionist is supported both by observational evidence—we find explanations for one level's functions by going to a deeper level, not the other way around—and recently by understanding some of the math behind it.

Having said that, I must address here a common misunderstanding about this layered structure of natural laws, namely that there seem to be examples contradicting it. Say you push a button that turns on a particle collider that collides two protons, which produces a Higgs boson. In that sequence, wasn't it your decision—i.e., an upper-level function—that caused an event on much shorter distances, hence violating the idea of this neatly ordered structure? Another common example is that of computer algorithms that switch transistors on and off while processing information. Isn't it the algorithm that you programmed—upper-level function—that controls the electrons? It isn't hard to come up with an abundance of similar examples.

The misunderstanding in these cases is always the same. Just because it is useful to describe certain properties or behaviors of a system (you, a computer algorithm) in macroscopic terms (motives, computer code) doesn't mean the macroscopic description is more fundamental. It isn't. You could full well describe a computer, including its algorithms, in terms of neutrons, protons, and electrons. It would be a totally useless description, of course.

But if you wanted to prove reductionism false, you'd have to show that describing a system in macroscopic terms results in *different* predictions than those you'd get from its microscopic description (and then do an experiment that demonstrates that the predictions from the microscopic description are wrong). No one has managed to do that. Again, it's not because that wouldn't have been possible. Maybe you can try to conceive of a world in which the behavior of atoms derives from that of the planets instead of the other way round, but for all we can tell, it's just not the case.

To make sense of this tower of theories, note that the function of a composite object does not derive merely from its constituents. One also has to know the interactions of the constituents and the correlations between them, i.e., one needs the full microscopic information. Quantum entanglement, in particular, is really a type of correlation— it links particles together—but even though it can span macroscopic distances, it's still a property defined on the fundamental level. We will talk more about entanglement later, but let us note for now that it doesn't contradict reductionism.

In summary, according to the best current evidence, the world is reductionist: the behavior of large composite objects derives from the behavior of their constituents, but we have no idea why the laws of nature are that way. Why is it that the details from short distances do not matter over long distances? Why doesn't the behavior of protons and neutrons inside atoms matter for the orbits of planets? How come what quarks and gluons do inside protons doesn't affect the efficiency of drugs? Physicists have a name for this disconnect—the *decoupling of scales*—but no explanation. Maybe there isn't one. The world has to be some way and not another, and so we will always be left with unanswered *why* questions. Or maybe this particular *why* question tells us we're missing an overarching principle that connects the different layers.

One Bit at a Time

If you're at all like me, you probably think of yourself as a physically compact, localized object, feet at one end, head at the other. That intuitive self-image, however, isn't rooted in reality.

Our bodies' physical composition constantly changes. We swap some of the particles we're made of for new ones each time we breathe, drink, or eat. After all, that's how we grew to this size to begin with. Throughout our lives, we repurpose atoms that previously belonged to other animals, plants, soil, or bacteria, atoms that were created in the Big Bang or by stellar fusion. A carbon-dating study in 2005 found that the average cell in the adult human body is only seven years old. Though some cells stay with us pretty much our whole life, skin cells are on average replaced every two weeks, and others (like red blood cells) are replaced every couple of months.

We are, hence, physically less like the compact object our self-image suggests, and more like the ship of Theseus. In this 2,500-year-old mind twister, a ship of the Greek hero Theseus is put up in a museum. As time passes, parts of the ship begin to crumble or rot away, and bit by bit they are replaced with newer parts. A rope here, a plank here, a mast there. Eventually, none of the original pieces is left. "Is it still the same ship?" the Greek philosophers wondered. From this ancient debate comes the saying "No man ever steps in the same river twice, for it's not the same river and he's not the same man," which is usually attributed to Heraclitus.[a]

As so often, the answer depends on how you define the terms in the question. What do you mean by *the ship* and what do you mean by *the same*? It's only once you have defined these expressions that you'll

---

[a]Though Heraclitus didn't actually write that. Is a quote in which word after word has been replaced until none of the original words remain still the same quote? The answer is left as an exercise for the reader.

be able to answer the question—and there are many different answers. Don't worry; I have no intention to roll up 2,500 years of
philosophy—I'll get back to physics in a moment—but credit where
credit is due: the old Greeks realized long ago that an object's constituents aren't the only thing that's relevant about it. Even after you
have exchanged all the pieces of the ship, its construction plan—the
information you need to build it—remains the same. Indeed, you
could define information as what doesn't change about the ship when
you replace its parts.

It's similar for humans. Humans are made of particles, and the
behavior of those particles determines our behavior. But that reduction
isn't what makes humans—or any complex structure—interesting.
What makes them interesting are the emergent higher-level properties: Humans walk, talk, and write books. Some of them reproduce.
Others fly to the moon. Jars of chemicals don't do that. The relevant
property of humans is not our constituents. It's the way the constituents are arranged; it's the information you need to build a human, the
information that tells you what it can do.

I don't mean just your genetic code, for your genes alone aren't
sufficient to define the person you are today. I mean all the necessary
details that specify the way each part of your body, each single molecule, interacts with any other. That includes the countless little (and
big) experiences that left marks in your brain, traces of the food
you've eaten and air you've breathed, legacies of past illnesses, scars,
and bruises. What makes you you is this entire arrangement. Your
you-ness, whatever exactly it is, emerges from the configuration of
the particles you are constituted of. For all we currently know, these
properties could emerge in different ways.

The Canadian scientist and philosopher Zenon Pylyshyn illustrated this nicely with a thought experiment in 1980. Imagine you
are going about your usual day-to-day thinking, maybe wondering

whether it's time for a coffee. Now, suppose someone takes away one of your neurons and replaces it with a silicon chip. The silicon chip is designed so it responds to input and output from the rest of your brain identically to the replaced neuron. The chip performs the exact same function that the neuron did previously, and it seamlessly connects to the other neurons. Does that change anything about your personality? Do you maybe suddenly forgo coffee and ask for tea? No. Why would it make a difference? After all, nothing changed about the way your brain processes information. Good. Then swap the next neuron with a chip. And the next. That way, one by one, your brain is replaced by silicon chips, until it's all silicon chips. Are you still the same person?

As with Theseus's ship, it depends on how you define *you* and *the same*. In some sense, you arguably aren't the same person, because you're now made of different physical components. Yet the physical components aren't what we care about. What we care about is the arrangement of the components. It's the functions they perform that make you interesting. In that sense, you haven't changed. You can still perform the same functions; you're still as interesting as you used to be.

But are you really the same? That's where physics becomes relevant. It's one thing to write that you can replace a neuron with a silicon chip without altering the brain's function whatsoever. Whether that's actually possible is another question entirely. The phrase *the same* in Pylyshyn's thought experiment implicitly assumes it is possible to replace neurons with chips, not just so the differences are unnoticeably small, but so there are no differences. This strong assumption is necessary for the argument to work. If I replaced one molecule of your coffee with one molecule of tea, it would taste the same to you; it's an unnoticeably small difference. But if I kept replacing molecule after molecule, eventually you would notice. A large number of unnoticeably small changes can eventually become a noticeably large change. How do we know that's not what's going on in the neuron replacement?

The obvious answer is that we don't know, because no one's done it. Still, we can ask what is possible according to all we know about physics. Is it possible to replace a neuron with something else so the substructure—silicon or carbon—makes no difference? Yes, it's possible because, as we discussed in the previous section, scales decouple. We can ignore the details on short distances for the emergent behavior on large distances. And this also means you can swap the short-distance physics for another one, neurons for chips, or yet something else, and it wouldn't make a difference, as long as the emergent behavior is the same.

Of course, as always, it could be that something is wrong with the theories we currently use and that this argument therefore fails for reasons we don't yet know of. The physicist and Nobel Prize winner Gerard 't Hooft, for example, has argued that the observations we attribute to quantum randomness are really due to noise that is as yet unaccounted for, arising from new phenomena at short distances. If that is so, the decoupling of scales could fail. Maybe 't Hooft is right, but so far his idea is pure speculation.

I should mention for completeness that it isn't entirely clear at present whether our brains are the sole home of our identity, but this complication doesn't matter for the argument. Studies have shown, for example, that at least some aspects of human cognition are *embodied*; that is, they rely on input from other body parts, such as the heart or guts. That may be bad news for people who've had their head frozen in the hope of being resurrected, but it isn't relevant to the question whether your constituents can be replaced by physically different parts. If swapping neurons in the brain doesn't entirely move your cognition to a silicon basis, then just imagine that the rest of your body is also being replaced.

The information that makes you you can be encoded in many different physical forms. The possibility that you might one day upload yourself to a computer and continue living a virtual life is arguably beyond present-day technology. It might sound entirely crazy, but it's compatible with all we currently know.

>> THE BRIEF ANSWER

You, I, and everything else are made of small constituent particles, and whatever large objects like us do is a consequence of what their many small constituents do. However, the characteristic features of a creature or object are the relations and interactions among their many constituent particles, not the particles themselves. For all we currently know, it is therefore possible to exchange the physical substrate of a being or object with something else. So long as this replacement maintains the characteristic relations and interactions, it should also maintain function, including consciousness and identity.

IS KNOWLEDGE PREDICTABLE?
An Interview with David Deutsch

It's here," the taxi driver says and points to a crumbling wall. Behind it thrives vegetation that hasn't seen a gardener for a long time. I am not at all sure this is the right address, but I figure it won't be a long walk from here. I pay the driver and step into a sunny autumn day. It's a quiet street in the outskirts of Oxford, where I am looking for David Deutsch.

Closer inspection of the house in front of me reveals it is the right address after all, and so I make my way to the door along a walkway infringed on by plants. The door is surrounded by cobwebs and could use new paint. I ring the bell. It doesn't take long for David to open the door.

Even for someone as nearly face-blind as I, David Deutsch is easy to recognize. His eyes seem too big for his sharp nose and lean face, and his hair, like that of most British men, is too long. He welcomes me with a big smile and asks me in. The inside of his house, I see, isn't in any better condition than the outside, but as a mother of two children in primary school, I am practiced in carefully treading around stacks of toys, books, and unidentifiable craftworks. That skill comes in handy now.

David leads me to what I believe must be the living room. It contains a couch opposite a huge flat screen on a desk, some folding chairs and bookshelves (from one of which I see Charles W. Misner, Kip S. Thorne, and John Wheeler's *Gravitation* greeting me), gardening tools, lots of boxes, cables, various computer accessories, a blue mini-trampoline, and a bright-red Japanese massage chair. The massage chair, David enthuses, is new, and he sets out to demonstrate its various functions. It takes all my willpower not to ask him for a mop and a vacuum cleaner. Instead, I accept a glass of water and look for my notepad.

David is best known for his seminal contributions to quantum computing for which, in 2017, he received the Dirac Medal of the International Centre for Theoretical Physics, adding it to a long list of awards and honors. But I am not here to talk about quantum computing. I am here because I have been most impressed by David's popular science books, *The Fabric of Reality* and *The Beginning of Infinity*. It isn't only that David is exceedingly careful in laying out his rationale for thinking about what he is thinking about. It's also that he strikes me as a scientist who is way ahead of his time, interested not so much in the technologies of today as in the question how scientific knowledge grows, how it benefits our societies, and what knowledge is in the first place. David seems the right person to consult about the limits of reductionism.

I begin by asking him, too, whether he is religious. He answers with a straightforward *no*. He doesn't seem to have anything to add, so I move on to reductionism. "From a particle physics point of view, everything is made up of small particles, and in principle all the rest derives from it. Do you subscribe to this idea or do you think there are some things that cannot be reduced to their parts?"

"I don't subscribe to reductionism as a philosophy," David says.

"That is, I don't subscribe to the idea that the only true explanations are reductive ones."

"Just to be clear, what type of reductionism do you mean?" I ask. For most purposes, the distinction doesn't matter, but there are two types of reductionism. One is *theory reductionism*—levels of theories whose higher levels can be derived from the deeper ones, as we discussed in the previous chapter. The other one is *ontological reductionism*, which means that we get better explanations by physically going to smaller and smaller scales. The distinction usually doesn't matter, though, because they've historically gone hand in hand.

"I think both are false as philosophical principles," David answers. "But it so happens that some of the best theories of all time have been reductionist in both senses. For example, the periodic table. This was one of the explanatory triumphs of the nineteenth century, which linked all sorts of explanatory ideas, including the idea, resurrected from antiquity, that matter can't be infinitely subdivided. And like all solutions, this raised new problems. If the atoms can't be subdivided, how come they have different properties? And how come these properties are regular? This suggests that there has to be an underlying structure. And that's also how I see modern particle physics. It's like chemistry was in the nineteenth century. Maybe unlike chemistry in the nineteenth century, it has a taint of this reductionist philosophy that only subdividing things into smaller things can ever be an explanation. . . . Ah, sorry, I've lost track. I got so excited about the periodic table, I have forgotten your question!"

"You were saying that some of the best theories that we have are reductionist in both ways."

"Ah, yes"—David picks up his lost thread—"but some of them are not. For example, the theory of universal computation, which says that all the laws of physics are, say, Turing computable. In terms of physics, that means that there's a possible physical object, like this computer, such that the set of all possible motions of this thing corresponds one-to-one, in some approximation, to the set of all possible motions of *everything*."

He gestures at his laptop and continues. "Now, that's a powerful statement about the universe, and most conceivable laws of physics would not satisfy it. We think that the actual laws do satisfy it. And yet this principle refers to a thing—the universal computer—that is highly composite and highly complex. So if this is a fundamental principle that all laws are Turing computable, then this law is not reductive, and reductionism is false right there. It's saying that a particular high-level object has fundamental properties. And I think there is scope for future laws of that kind. Of course we will accept them only if they are good explanations. But that they are not reductive is not a criticism in my view."

He adds, "Similarly, if a law is reductive, this is also not a criticism. Some people are the opposite: they are holists. They think that a reductionist explanation can never be fundamental. I think that's false as well."

"You said you have this computer, so you have a higher-level object that has fundamental properties. But what exactly do you mean by *fundamental* here?"

"I mean that they are principles that we think are deep, universal truths about the world and not just accidentally true," David says. "Take, for example, the statement that there exists a solar system with eight planets, and the first three are rocky. We know that is true, because we live in one, but we don't think that's a fundamental statement. But the law of the conservation of energy, we think, is a deeper truth. And because it's deeper, it's a guide to future theories. When we are trying to write down new laws which fundamental particles may obey, we typically write down laws that obey energy conservation. We treat it as a principle that doesn't need to be explained by anything else."

"So it's a fundamental principle, but it's not reductionist, because it applies to everything?"

"We don't derive energy conservation from other laws," David explains. "We derive other laws from *it*. Well, of course, it may be false. But for it to be false, you need to have an explanation under which this could be false. For example, there are some understandings of general relativity in which energy is not conserved. And if that turned out to be correct, you would abandon that principle. It may happen, because general relativity is not totally satisfactory, as you know; we need a theory of quantum gravity."

I suggest, "Maybe the reason we don't have a theory of quantum gravity is that we're too tied to the idea that the more fundamental laws can be found on shorter distances. Is going to shorter distances maybe the wrong thing to do?"

"Yes, indeed!" David agrees. "As you know, I have this theory, constructor theory, in which the fundamental laws are not reductive. It's a very crude theory at present, but you have to stick your neck out at first. In constructor theory, the low-level, microscopic laws are all emergent properties of the high-level laws, not vice versa."

"Have you ever heard of something called the *causal exclusion principle*?" I ask.

"No."

"It seems to contradict what you just said," I explain. "So, in particle physics, we have this idea that if we combine small things to large things, then the laws for the small things tell us what the large things do. And we use the mathematical framework of effective field theory for this. This tells us that we do have a law for the macroscopic things already. The causal exclusion principle then says that, since we already have a law for macroscopic things, then any other macroscopic law is either derivable from the one we already have, or one of them must be wrong."

David replies, "I have no quarrel with the idea that the dynamical laws for macroscopic objects are deterministic and can be derived from the microscopic laws. But this doesn't imply that that's a good explanation."

I am still not sure I understand this entirely. "So constructor theory is not reductionist in the sense that the explanations don't start with the small scales?"

"Yes," David says. "As an example, let us just suppose that under constructor theory one of the fundamental laws says that there exist universal computers. In fact, let us say there exist universal computers with arbitrarily large memories. This one"—he points to his laptop again—"is an approximation to that, but in the future there will be ones with larger memories, and in the unlimited future, there'll be computers with unlimited memory. And suppose, just for the sake of the argument, that that really is one of the fundamental laws, but that other fundamental laws are reductive, such as quantum mechanics and elementary particle interactions, and so on.

"Well, then the existence of universal computers plus the microscopic dynamical laws translates into a statement about the initial state of the universe. But it translates into it in a highly intractable way. There would be no way of actually calculating all the properties the initial state must have, apart from saying it's such as to produce that end result: universal computers. Some people would rule this out; they would say it's a teleological theory. But it's not any old teleological theory. We have to explain why the universe is so that it has computers at all. The existence of even the type of computer we use today makes the laws of physics extremely unusual. It's as unusual as the existence of chemical elements. It's a feature of the world that we see and that we haven't explained."

I say, "But, of course, putting the thing that you want to explain into your theory doesn't explain it. If you say the universe is such that it goes on to produce computers, that doesn't explain anything."

"Right," David says. "You might as well say that the reason we are sitting here, and you are looking skeptical of what I say, is that you are going to write a book which will say, 'And I was skeptical of what he said.' And that you will write this in your book is the reason why you are skeptical now. That is the same argument, yet it explains nothing. I had to put the example with the computer this way because we don't yet have that theory which would explain it."

"OK," I say. "So you mean there *could* be a theory that has the property that it *would* go on to produce universal computers with unlimited memory and so on, but you don't know what that theory is."

"Yes," David agrees. "But what we do have of constructor theory is friendly to that kind of thing. It's not silly to imagine that an explanatory theory of that type exists."

Coming back to the question whether the future is determined, I ask, "You said you have no problem with the dynamical laws being deterministic. Would you say that for that reason everything is deterministic—not just computers but also human consciousness and behavior and so on?"

David says, "Yes, deterministic, in that as a matter of logic, the state at one time is determined by the state at any other time plus dynamical laws. But it may be that the later time is *explained* by the earlier time and not vice versa."

"But just because it's deterministic doesn't mean it's predictable," I say. "Do you mean that it's actually predictable?"

"No," he says. "For three reasons. First, in quantum mechanics, we cannot measure the state perfectly accurately. Therefore, even if we knew what the evolution of each state would be, we don't know what the actual state is, because it can't be measured.

"Second, there is chaos. Now, quantum mechanics doesn't have chaos,[a] but things like computers and brains do have it at the level that they work, so that means that changing even one bit in this computer will drastically change what it will do in the future. And because we can't measure our state of mind to anywhere near perfect accuracy, we are unpredictable.

"And there's a third reason, that's the most important, and that is that one can't predict the future growth of knowledge. No theory can be so good that it predicts the content of its own successor. Imagine you put a person into a glass sphere and do not allow interactions with the outside world and so on. You may think that in principle you can then predict everything that that person will do. But that's an illusion. Because if the person comes up with new ideas, such as a new law of physics, then there's no way you could have known that when you started the experiment. And if your computer works out what he'll do (say the computer works out in one day what he does in seven days), then you already have the new law of physics before he does, and the computation that the computer performed is actually a person: it is essentially him. So in order to calculate what he would do in the future, you really had to take him out of the sealed glass compartment and put him in a computer and run him in a virtual form. Oh, I should say that I think running someone in virtual form is exactly the same as running them in real form. Thinking is just a computation."

"So you are saying it would no longer be a prediction, because then you'd have the real thing in your computer already?"

---

[a] *Quantum chaos* relies on a nonstandard definition of the word *chaos* and is not a contradiction of what David said.

"Yes," David says. "We can't predict the future growth of knowledge, because if we could, we would have it before the moment that we are trying to predict. It's a feature of knowledge that it leads to unpredictability, even in a deterministic system."

"So to come back to what we talked about earlier," I say, "if we insist on reducing laws to more fundamental laws by going to smaller scales, then the growth of knowledge remains unexplained?"

"Among many other things, yes," David says. "Atomic theory was thought of without evidence. The problem that they had in antiquity was that if the world was a continuum, to get from *A* to *B* you need to pass through an infinite number of points. And if it's not a continuum, how do you get from one discrete point to the next? Both seem impossible. The theory of atoms was developed because they tried to find a way out. And that might seem so esoteric that it would have no practical implication at all. But [ideas like this] were the things that led to everything good. And this is my view of the role of particle physics, reductionism, and holism. They should all be subordinated to the task of explaining the world."

And I was skeptical of what he said.

>> *THE BRIEF ANSWER*

If you could predict the growth of knowledge, your knowledge wouldn't grow.

Chapter 5

DO COPIES OF
US EXIST?

Many Worlds

Popular-science news about quantum mechanics is to me as baffling as it is frustrating. Hand me an equation, and I can deal with it. But if you tell me that quantum mechanics allows one to separate a cat from its grin or that an experiment shows "an irreconcilable mismatch between the friends and the Wigners," I'll back out of the room quietly before anyone demands I explain this mess. I have suffered through countless well-intended introductions to quantum mechanics featuring quantum shoes, quantum coins, quantum boxes, and entire zoos of quantum animals that went in and out of those boxes. If you actually understand those explanations, I salute you, because if I hadn't known already how quantum mechanics works, I still wouldn't know.

I am telling you this not to undermine your fun with quantum shoes, but so you understand where I am coming from. I am very much a math person, and personally I don't see the need to translate math into everyday language. Abstract mathematical structures, I think, are best dealt with on their own terms. They don't need to be interpreted, and they don't need to make intuitive sense. They don't have to be "like" anything else—because in most cases, they are not. The whole reason we use this math is that there isn't anything else like it.

To me, quantum mechanics is the prime example for what can go wrong with using intuitive language for abstract math. Take superpositions. In quantum mechanics, we put initial states into the Schrödinger equation to calculate how they change in time. The Schrödinger equation has the property that if we have solved it for two different initial states, then the sum of those solutions, each multiplied by arbitrary numbers, is also a solution to the equation.[a] And that's what a superposition is: A sum. That's it. No, I'm not joking. Entangled states are a specific type of superposition. Yes, they're also sums. So where did all the fabled weirdness go?

The weirdness appears only if you try to express the math verbally. If one of the states that solved the Schrödinger equation describes a particle moving to the right, and the other one a particle moving to the left, then what's the sum of those? It has become common to use the phrase "the particle goes in both directions at the same time." Does that adequately describe what a superposition is? In all honesty, I don't know. I'd prefer to leave it at "it's a superposition."

Of course, I understand the need to express mathematics in words to make it accessible, which is why I myself have used metaphors for superpositions when I don't have time or space to explain the details. And I will do the same here: omit the math and try to give you an idea of what it all means. But I want you to know that much of the supposed weirdness of quantum mechanics just comes from forcing it into everyday language. There are no exact metaphors, not for quantum

---

[a]That just means it's a linear equation, as opposed to nonlinear equations that we have in chaotic systems and also in general relativity.

mechanics and not for anything else, because if they were exact, they wouldn't be metaphors.

It doesn't help that calling quantum mechanics *strange, weird,* or *spooky* makes for catchy headlines, and thus popular-science outlets use these words a little too frequently and a little too gleefully. I agree with Philip Ball that, at more than a hundred years of age, quantum mechanics should move "beyond weird." Having said that, let us look at what quantum mechanics tells us about afterlife.

o o o

Without quantum mechanics, the laws of nature are deterministic. To recap, this means if you have an initial state, you can calculate unambiguously what happens at any later time. Say you drop a pen and it falls to the ground. If you could measure exactly where and from which position the pen started, and if you knew the exact locations and motions of all the air molecules around it, you could calculate when and how the pen would land.

Of course, we can't measure the positions of all air molecules exactly, and even if we could, using them to predict the outcome would be unfeasible. But in principle, without quantum mechanics, any uncertainty about the outcome arises merely from our lack of knowledge about the initial conditions. We call these types of nonquantum theories **classical**.

Quantum mechanics works differently. In quantum mechanics, we describe everything by wave functions. There is a wave function for electrons and one for photons, but there are also wave functions for grapefruits, brains, and even one for the whole universe. These wave functions evolve partly deterministically, but every once in a while, when a measurement takes place, they make indeterministic jumps.

These jumps are not entirely unpredictable—we can predict the probability that they happen and the probability for their outcome—but they have an element that is fundamentally random. This uncertainty in the outcome of a measurement in quantum mechanics is not due to our lack of knowledge about the initial conditions; according to quantum mechanics, that's just how it is.

This unpredictable randomness of quantum mechanics is not confined to subatomic scales, so it's not like you can just wave it away as an irrelevant quirk of nature that scientists occasionally see in their laboratories. Exactly because the outcome of a measurement is what's unpredictable, the randomness manifests itself for macroscopic objects like you and me.

Suppose that an experimenter observing a flash on a screen goes home when the particle appears on the left side and stays in the lab when it appears on the right side. Maybe that decides whether or not she gets into an accident on the highway. The randomness of a single quantum event can change her whole life. And this doesn't happen only in the laboratory. If a cosmic ray hits living tissue, the damage to the genetic code, for example, ultimately comes down to quantum indeterminism.

But while quantum mechanics is an extremely successful theory, just what its mathematics means has been disputed since the development of the theory at the beginning of the twentieth century. Some have argued that nature can't be fundamentally random—like Einstein, who claimed that "God does not play dice"—and that quantum mechanics is just incomplete. Others, like one of the founders of quantum mechanics, Niels Bohr, have claimed that we just need to get over old-fashioned ideas of determinism.

Most physicists today ignore the entire debate and deal with quantum mechanics as a tool that makes predictions and that one shouldn't overthink. This "shut up and calculate" attitude is the pragmatic way to go about it. It has led to great progress, so it shouldn't be laughed off. However, many researchers who work in the foundations of physics feel that ignoring the problems of quantum mechanics is a mistake because we'd learn more from resolving them.

To understand the problem with quantum mechanics, recall that in Einstein's theory of special relativity, nothing can happen faster than the speed of light. Yet in quantum mechanics, the moment you make a measurement, probabilities change, instantaneously and everywhere. This update of the wave function is **nonlocal**. It is, as Einstein put it, a "spooky action at a distance." Alas, it turns out that in that process of measurement, no information is submitted faster than the speed of light. Indeed, one can mathematically prove it's impossible to send information faster than the speed of light with quantum mechanics. So it's not as though there's something concretely wrong with the theory. It just feels wrong.

Researchers have proposed various ways to deal with this situation. Some argue that quantum mechanics is just not the right theory and has to be replaced with something better. This is a possibility I have worked on myself, but because it's both speculative and somewhat off topic, I don't want to go into this here. For the purposes of this book, I will stick with what is widely considered the accepted status of research.

If you don't want to actually change quantum mechanics, you can try to interpret the mathematics differently and hope that then it makes more sense. There are a number of such interpretations. For example, there's the interpretation proposed by Niels Bohr, according to which the wave function just shouldn't be considered real. It's a device to make predictions for measurements, Bohr said, but if you are not making a measurement, it is meaningless to ask what is really happening. This is now often called the *Copenhagen interpretation* or just the *standard interpretation*, because it's the most commonly taught one.

Needless to say, a lot of physicists don't like being told they're not supposed to ask questions, so they have tried to find other, more intuitive ways to make sense of the math. One alternative interpretation was pursued by David Bohm, and is known today as *Bohmian mechanics*.

Bohm reformulated the equations of quantum mechanics so they look more similar to those of classical mechanics. In Bohm's equations, the wave function is still there, but now it describes a field that "guides" particles. The indeterminism in the measurement outcomes, according to Bohm's interpretation, is due to a lack of knowledge, as it is in classical physics. Alas, Bohmian mechanics also has it that you can never remedy that lack of knowledge. In the end, the outcome is exactly the same as in the Copenhagen interpretation. Bohm's interpretation never became very popular, but it still has its followers today.

Yet another way of interpreting quantum math was pioneered by Hugh Everett and further developed by Bryce DeWitt. They argued that one should just get rid of the measurement update and thereby return to a deterministic evolution. In the many-worlds interpretation, each possible measurement outcome happens, but it happens in its own universe. If you think back to the particle that had a fifty-fifty chance of hitting the screen on the left side or the right side, then in the many-worlds interpretation, it will hit the screen on the right side in one universe and on the left side in another universe. And after it does that, these two universes will remain separated forever—they will evolve in their own *branches*, as they are often called.

Before we can move on, I have to sort out a common misunderstanding with the many-worlds interpretation. One way to explain how quantum mechanics works, one you may have come across, is that a particle takes every possible path from its initial to its final place. For example, if one points a laser beam at a screen with two slits (the famous *double slit*), then each particle in the laser beam goes through both slits. It's not that one particle goes through the left and the next one through the right slit; each goes through both slits (figure 8).

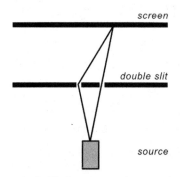

Figure 8: *One can interpret the double-slit experiment by saying that the particle takes all paths at the same time. Shown are the two most likely paths to one particular place on the screen.*

This, again, is an interpretation of the mathematics, originally proposed by Feynman, called the *path-integral approach*. Mathematically, you then have to sum over all the possible paths to calculate the probability of the particle to go to one particular place. To make a long story short, the result is the same as in the original formulation with the Schrödinger equation, but physicists like using path integrals because the approach can be generalized to more-difficult situations.

One can interpret the path integral as telling us that the particle takes each path in a different universe. Personally, I find this a rather meaningless statement—there isn't anything in the math that says these paths are in different universes—but it isn't wrong either. And I am all for different ways of looking at math because they can lead to new insights. So, OK.

However, these different paths—or universes, if you wish—in the usual path integral are present only *before* the measurement. In the many-worlds interpretation, in contrast, the different universes still exist *after* the measurement. So just because we can reformulate quantum mechanics in terms of a path integral doesn't mean the many-worlds interpretation is correct. These are two different things.

The key feature of the many-worlds interpretation is that each time a quantum measurement happens, the universe splits, creating what's commonly called a *multiverse*.[a] And because we've seen earlier that (with some apologies about the abuse of terminology) even interactions with air or the cosmic microwave background can cause a measurement, that creates a lot of universes really quickly. It also makes a lot of physicists uncomfortable really quickly.

The problem with the idea is that, well, no one's ever seen a universe splitting. According to the many-worlds interpretation, that's because detectors and their generalization—like you and I—split together with the universes. What determines which universe you go into? Ah, you supposedly go into all of them. Because that isn't what we experience, the many-worlds interpretation requires further assumptions (besides the Schrödinger equation) that specify how to calculate the probability of going into one particular universe. This brings indeterminism in again through the back door.

I'll spare you the mathematical details because they don't really matter. The upshot is that you need to add sufficiently many assumptions to reproduce the predictions of what was formerly called the measurement update. Because—you know what?—it was there for a reason, the reason being that it's necessary to describe what we observe, and if you just throw it away, then the theory simply does not give the right predictions. We do not, as a matter of fact, observe all possible outcomes of an experiment.

This means that as far as calculations are concerned, the many-worlds interpretation makes exactly the same predictions as quantum

---

[a]Yes, it would etymologically have made more sense to call the multiverse the *universe* and then maybe refer to what we previously called the universe as a *sub-universe*. But language rarely follows the rules of logic.

mechanics in the standard interpretation, from equivalent, yet differently expressed, assumptions. The major difference isn't in the math; the major difference is one of belief. Advocates of the many-worlds interpretation believe that all the other universes—the ones we don't observe—are as real as ours.

But in which sense are they real? Unobservable universes are by definition unnecessary to describe what we observe. Hence, assuming they are real is also unnecessary. Scientific theories should not contain unnecessary assumptions, for if we allow that, we would also have to allow the assumption that a god made the universe. Such superfluous assumptions aren't wrong. They're just ascientific. The assumption that the additional universes in the many-worlds interpretation are real is one such ascientific assumption.

I must stress that this doesn't mean that the parallel universes of the many-worlds idea are *not* real. It means that any statement about their reality is ascientific. It is something you can believe or disbelieve, but science tells you nothing—*can* tell you nothing—about what is correct. Conversely, this also means that the idea that there are infinitely many yous out there, somewhere, doesn't conflict with anything we know. It's a science-compatible belief system.

It does have a few odd consequences, though. For example, because deep down all our brain processes are quantum processes, for every decision you make, there'll be a universe in which you chose the other option. And just in case you aren't sure a decision was indeed based on a quantum effect, there's an app for this: the Universe Splitter sends a photon through a semitransparent mirror for you, and depending on whether it goes through or not, you can choose either pasta or chicken, accept or decline, take the red pill or the blue, all the while believing that a copy of you lives out the option you didn't choose.

That's quite something already, but the best example of odd multiverse consequences may be the idea of *quantum suicide*. Imagine you repeat an experiment in which a quantum process kills you with 50 percent probability. In the standard interpretation of quantum mechanics, the probability of your survival goes down by half in each repetition of the experiment. By the twentieth repetition, the probability that you are dead is 99.9999 percent.

However, according to the many-worlds interpretation, you don't die with 50 percent probability in each round. Instead, the first time you run the experiment, the universe splits into two universes, one in which you die and one in which you survive. In the second run, each of the two universes splits, so you have four. In two of them you died already in the first round, so the second experiment doesn't matter. Then there's one in which you survived the first round but die in the second, and one in which you survive both. Do the experiment again and all four universes split to make eight, and so on (see figure 9). After twenty rounds, you are still alive with 100 percent probability, but only in one of a million universes.

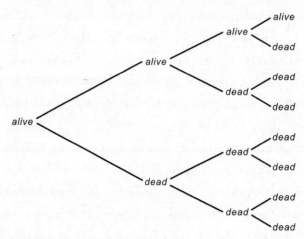

Figure 9: The many worlds of quantum suicide.

It gets better. Because every molecular process comes down to quantum mechanics, this means that whatever the cause of somebody's death, that person had a tiny, yet nonzero, probability of survival—quantum randomness makes it possible. There is always the chance that an illness will spontaneously go into remission, that cell damage will suddenly revert, or that a heart will start beating again after having given up. And even if the chance of that happening is minuscule, in the many-worlds interpretation, it will happen—for every one of us—in some branch of the multiverse.

This, of course, also means there's a branch in the multiverse in which dinosaurs still roam the world, Hitler was never born, and spray cheese was never invented. That's arguably not the branch we live in, so what are we to make of all that?

If you believe in the many-worlds interpretation, reasoning about probabilities in our universe becomes reasoning about numbers of branches in the multiverse. And because you can't go back in time to choose a different branch, those probabilities are relative to your present observation of the universe. For all practical purposes, the outcome is exactly the same: dinosaurs are extinct, the Second World War happened, and spray cheese is a thing. You may not die in all branches of the multiverse, but the probability of your survival (or that of the dinosaurs) decreases just the same as it does in the standard interpretation. That's why no one's attempted quantum suicide: it would decrease the number of universes in which they survive.

As far as observations are concerned, the many-worlds interpretation doesn't make any difference. But if you like to believe there are infinitely many copies of you living all possible alternative versions of your life, go ahead. That belief isn't in conflict with science.

The MultiWorse

The many-worlds interpretation is only one type of multiverse. There are a few others that have become popular in past decades.

One is an extension of the idea of inflation, the hypothetical phase of exponential expansion in the early universe. In this extension, *eternal inflation*, one invents a mechanism to create the initial state for inflation in our universe. The currently most popular way of doing this is to conjecture a multiverse in which Big Bangs happen all the time and everywhere. The universes created in the other Big Bangs could be similar to our own, or they could have different constants of nature, leading to entirely different physical laws.

That the constants of nature might change from one Big Bang to another comes from another multiverse idea: the *string-theory landscape*. String theorists originally hoped they'd be able to calculate the constants of nature. That didn't pan out, so now they argue that if they can't calculate the constants, that must mean all possible values exist somewhere in a multiverse.

You can combine all these different multiverses to one megamultiverse.

As in the many-worlds interpretation, in the other multiverses, the universes besides our own are also by construction unobservable.[a] And they are inhabited by even more copies of you, but these copies come about for a different reason: small variations in the initial state can lead to universes with histories almost, but not exactly, like the history of our own universe. Of course we don't actually know, and will never know, what initial states are actually possible in the

---

[a]There are some specific versions of multiverses that would have observable consequences: for example, that our universe might collide with, or become entangled with, another one. Alas, to the extent that those ideas were falsifiable, they have been falsified. It is thus moot to discuss them here.

multiverse because we can't gather observational evidence for them. This is pure conjecture.

The scientific status of these multiverse ideas is thus the same as that of the many-worlds interpretation. Assuming the reality of something unobservable is unnecessary to describe what we observe. Hence, assuming that these other universes are real is ascientific.

This isn't a particularly difficult argument, so I find it stunning that my physics colleagues can't seem to comprehend it. Inevitably, they will declare, "But then one would also have to argue that talking about the inside of black holes isn't scientific." But, no, the situation for black holes is entirely different. For one, you can (in principle!) totally observe what's inside a black hole. You just can't come back to tell us about it. More important, black holes evaporate, so their inside doesn't eternally remain disconnected from us. When a black hole evaporates, its horizon shrinks until it's entirely gone. If that weren't so, I would indeed question the scientific merit of talking about what's inside a black hole.

"But," they'll argue next, "because the speed of light is finite but the universe is only 13.7 billion years old, we can merely see part of the universe, even if it's infinitely large." Yet certainly I don't believe that the universe stops existing beyond the boundary of what we can currently observe.

Well, as I have explained so often, this isn't about what I believe or not; it's about what we can know or not. I am saying that what's beyond what we can observe is purely a matter of belief. Science doesn't say anything about whether it exists or doesn't exist. Hence, claiming it exists is ascientific, and so is claiming it doesn't exist. If you want to talk about it, fine, but don't pretend it's science. At that point, they're usually either confused or offended or both.

The reason I keep insisting physicists clean up their act and stop conflating belief with science is that their confusion is patently obvious to non-experts. Physicists from Brian Greene to Leonard Susskind to Brian Cox to Andrei Linde have publicly spoken about the multiverse as if it were best scientific practice. And because multiverse ideas attract a lot of media attention, this sheds a bad light on the capability of the scientific community to hold its members to high standards.

A prominent example for the damage that can result comes from 2016 Republican presidential candidate Ben Carson. Carson is a retired neurosurgeon who doesn't seem to know much about physics, but what he knows, he must have learned from multiverse enthusiasts. On September 22, 2015, Carson gave a speech at a Baptist school in Ohio, informing his audience that "science is not always correct." This is, of course, correct. But then he went on to justify his science skepticism by making fun of the multiverse:

> And then they go to the probability theory, and they say, "But if there's enough big bangs over a long enough period of time, one of them will be the perfect big bang and everything will be perfectly organized."

In an earlier speech, he cheerfully added, "I mean, you want to talk about fairy tales? This is amazing."

Now, it is clear from Carson's elaborations that he has misunderstood much of thermodynamics and cosmology, but, really, this isn't the point. I don't expect neurosurgeons to be experts in the foundations of physics, and I hope Carson's audience doesn't expect that either. The point is, he shows us what happens when scientists mix fact with fiction: non-experts throw out both together.

In his speech, Carson went on: "I then say to them, 'Look, I'm not going to criticize you. You have a lot more faith than I have. . . . I give you credit for that. But I'm not going to denigrate you because of your faith, and you shouldn't denigrate me for mine.'"

And I'm with him on that. No one should be denigrated for what they believe in. If you want to believe in the existence of infinitely many universes with infinitely many copies of yourself, some of whom are immortal, that's fine with me. But please don't pretend it's science.

Do We Live in a Computer Simulation?

I quite like the idea that we live in a computer simulation. It gives me hope that things will be better on the next level. This *simulation hypothesis*, as it's called, has been mostly ignored by physicists, but it enjoys a certain popularity among philosophers and people who like to think of themselves as intellectual. Evidently, it's more appealing the less you understand physics.

The simulation hypothesis is most strongly associated with the philosopher Nick Bostrom, who has argued that (given certain assumptions I will come to in a moment) pure logic forces us to conclude that we are simulated. Elon Musk is among those who have bought into it. "It's most likely we're in a simulation," he said. And even Neil deGrasse Tyson gave the simulation hypothesis "better than 50-50 odds" of being correct.

The simulation hypothesis annoys me, but not because I'm afraid people will actually believe it. Most people understand that the idea lacks scientific rigor. No, the simulation hypothesis annoys me because it intrudes on the terrain of physicists. It's a bold claim about the laws of nature that doesn't pay any attention to what we know about the laws of nature.

Loosely speaking, the simulation hypothesis has it that everything we experience was coded by an intelligent being, and we are part of that computer code. The opinion that we live in some kind of computation in and by itself is not an outrageous claim. For all we currently know, the laws of nature are mathematical, so you could say the universe is really just computing those laws. You may find this terminology a little weird, and I would agree, but it's not controversial. The controversial bit about the simulation hypothesis is that it assumes there is another level of reality where some being or some thing controls what we believe are the laws of nature, or even interferes with those laws.

The belief in an omniscient being that can interfere with the laws of nature, but that for some reason remains hidden from us, is a common element of monotheistic religions. The difference is that those who believe in the simulation hypothesis argue that they have arrived at their belief by reason. Their line of argumentation usually closely follows Nick Bostrom's argument, which, in a nutshell, goes like this: if there are (a) many civilizations, and these civilizations (b) build computers that run simulations of conscious beings, then (c) there are many more simulated conscious beings than real ones, so you are likely to live in a simulation.

First of all, it could be that one or both of the premises is wrong. Maybe there aren't any other civilizations, or they aren't interested in simulations. That wouldn't make the argument wrong, of course; it would just mean that the conclusion can't be drawn. But I will leave aside the possibility that one of the premises is wrong, because I don't really think we have good evidence for one side or the other.

The point I have seen people criticize most frequently about Bostrom's argument is that he just assumes it is possible to simulate humanlike consciousness. We don't actually know that this is possible. However, in this case, it would require an explanation to assume that it is not possible. That's because, for all we currently know, consciousness is simply a property of certain systems that process large amounts of information. It doesn't really matter exactly what physical basis this information processing is based on. It could be neurons or transistors, or it could be transistors believing they are neurons. I don't think simulating consciousness is the problematic part.

The problematic part of Bostrom's argument is that he assumes it is possible to reproduce all our observations not using the natural laws that physicists have confirmed to extremely high precision but using a different, underlying algorithm, which the programmer is running. I don't think that's what Bostrom meant to do, but it's what he did. He implicitly claimed it's easy to reproduce the foundations of physics with something else. This is the problematic part of the argument.

To begin with, quantum mechanics features phenomena that are not computable with a conventional computer in finite time. At the very least, therefore, one would need a quantum computer to run the simulation—that is, a computer that works with quantum bits, or *q-bits* for short, that are superpositions of two states (say, 0 and 1).

But nobody yet knows how to reproduce general relativity and the standard model of particle physics from a computer algorithm running on any type of machine. Waving your hands and yelling "quantum computer" doesn't help. You can *approximate* the laws we know with a computer simulation—we do this all the time—but if that were how nature actually worked, we could see the difference. Indeed, physicists have looked for signs that natural laws really proceed step by step, like a computer code, but their search has come up empty-handed. It's possible to tell the difference because all known attempts to algorithmically reproduce natural laws are incompatible with the full symmetries of Einstein's theories of special and general relativity. It's not easy to outdo Einstein.

This problem exists regardless of what the laws are in the higher-level reality from which the programmer supposedly simulates us. We don't know any kind of algorithm that would give us the laws we observe, regardless of what that algorithm is running on. If we knew, we'd have found a theory of everything.

A second issue with Bostrom's argument is that, for it to work, a civilization needs to be able to simulate a lot of conscious beings, and these conscious beings will themselves try to simulate conscious beings, and so on. While you can imagine simulating a single brain with its inputs only, in this case the conclusion we are likely to live in a simulation because there are more simulated than real brains wouldn't work. You actually need a lot of brains. But this means you have to compress the information we think the universe contains because otherwise your simulations will run out of disk space quickly. Bostrom therefore has to assume it's somehow possible to not care much about the details in some parts of the world where no one is currently looking, and just fill them in in case someone looks.

Again, though, he doesn't explain how this is supposed to work. What kind of computer code can actually do that? What algorithm can identify conscious subsystems and their intention and then quickly fill in the required information without ever producing an observable inconsistency? That's a much more difficult issue than Bostrom seems to appreciate. Not only does it assume that consciousness is computationally reducible, for otherwise you couldn't predict where someone is about to look before they look, but also you can in general not just throw away physical processes on short distances and still get the long distances right.

Global climate models are an excellent example. We don't currently have the computational capacity to resolve distances below something like 10 kilometers (6¼ miles) or so. But you can't just discard all the physics below this scale. This is a nonlinear system, so the details from short scales leave a mark on large scales—butterflies causing tornadoes and so on. If you can't compute the short-distance physics, you have to at least suitably replace it with something. Getting this right even approximately is a big headache. And the only reason climate scientists do get it approximately right is that they have observations they can use to check whether their approximations work. If you have only a simulation, like the programmer in the simulation hypothesis, you can't do that.

That's my issue with the simulation hypothesis. Those who believe it make big assumptions, maybe unknowingly, about what natural laws can be reproduced with computer simulations, and they don't explain how this is supposed to work. But finding alternative explanations that match all our observations to high precision is really difficult. I should know—it's what we do in the foundations of physics.

Maybe you're now rolling your eyes because, come on, let the nerds have some fun, right? And, sure, some part of this conversation is just intellectual entertainment. But I don't think popularizing the simulation hypothesis is entirely innocent fun. It's mixing science with religion, which is generally a bad idea, and, really, I think we have better things to worry about than that someone might pull the plug on us.

In summary, the simulation hypothesis isn't a serious scientific argument. This doesn't mean it's wrong, but it means you'd have to believe it because you have faith, not because you have logic on your side.

>> THE BRIEF ANSWER

The idea that copies of us exist in the multiverse is not scientific, because such copies are both unobservable and unnecessary to explain what we *can* observe. Multiverse theories have been promoted by physicists who believe that mathematics is real, as opposed to being a tool to describe reality. You are, therefore, welcome to believe that copies of you exist, if you want, but there is no evidence this is actually correct. The hypothesis that our universe is a computer simulation does not meet the current scientific standard.

Chapter 6

HAS PHYSICS RULED
OUT FREE WILL?

A Quagmire of Evasion

The major problem with discussions about free will is that philosophers have put forward a heap of definitions that have nothing to do with what nonphilosophers think *free will* means. I am tempted to write "normal people" as opposed to "philosophers," but maybe that's a little uncharitable. And I don't want to be uncharitable. Certainly not.

For this reason, let me begin by stating the problem without using the term *free will*. The currently established laws of nature are deterministic with a random element from quantum mechanics. This means the future is fixed, except for occasional quantum events that we cannot influence. Chaos theory changes nothing about this. Chaotic laws are still deterministic; they are just difficult to predict, because what happens depends very sensitively on the initial conditions (butterfly flaps and all that).

Our life is thus not, in Jorge Luis Borges's words, a "garden of forking paths" where each path corresponds to a possible future and it is up to us which path becomes reality (figure 10). The laws of nature just don't work that way. For the most part, there is really only one path, because quantum effects rarely manifest themselves macroscopically. What you do today follows from the state of the universe yesterday, which follows from the state of the universe last Wednesday, and so on, all the way back to the Big Bang.

But sometimes random quantum events do make a big difference in our lives. Remember the researcher who might get into a highway accident depending on where a particle appeared on her screen? The paths do fork every once in a while, but we have no say in it. Quantum events are fundamentally random and not influenced by anything, certainly not by our thoughts.

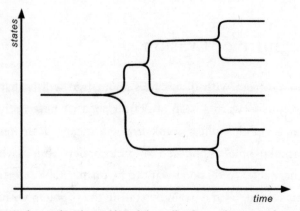

Figure 10: Forking paths. The trouble with free will is that we don't get to choose what happens at the forks.

As promised, I didn't use the term *free will* to lay out the situation. Let's then discuss what it means that the future is fixed except for occasional quantum events that we cannot influence.

Personally, I would just say this means free will does not exist and put the case to rest. I feel encouraged in that because free will itself is an inconsistent idea, as a lot of people wiser than I am have pointed out before. For your will to be free, it shouldn't be caused by anything else. But if it wasn't caused by anything—if it's an "uncaused cause," as Friedrich Nietzsche put it—then it wasn't caused by you, regardless of just what you mean by *you*. As Nietzsche summed it up, it's "the best self-contradiction that has been conceived so far." I'm with Nietzsche.

This is how I think about what's going on instead: our brains perform computations on input, following equations that act on an initial state. Whether these computations are algorithmic is an open question that we'll come to later, but there's no magic juice in our neocortex that puts us above the laws of nature. All we're doing is evaluating what are the best decisions to make given the limited information we have. A decision is the result of our evaluation; it does not require anything beyond the laws of nature. My phone makes decisions each time it calculates what notifications to put on the lock screen; clearly, making decisions does not necessitate free will.

We could have long discussions about just what it means that a decision is "the best," but that's not a question for physics, so let us leave it aside. Point is, we are evaluating input and trying to optimize our lives using some criteria that are partly learned and partly hard-wired in our brains. No more, and no less. And none of this conclusion depends on neurobiology. It is still unclear how much of our decision-making is conscious and how much is influenced by subconscious brain processes, but just how the division into conscious and subconscious works is irrelevant for the question whether the outcome was determined.

If free will doesn't make sense, why, then, do many people feel it describes how they go about their evaluations? Because we don't know the result of our thinking before we are done; otherwise, we wouldn't have to do the thinking. As Ludwig Wittgenstein put it,

"The freedom of the will consists in the fact that future actions cannot be known now." His *Tractatus* is now a century old, so it's not like this is breaking news.

Case settled? Of course not.

Because one can certainly go and define *something* and then call it free will. This is the philosophy of *compatibilism*, which has it that free will is compatible with determinism, never mind that—just to remind you—the future is fixed except for the occasional quantum event that we cannot influence. Among philosophers, compatibilists are currently the dominant camp. In a 2009 survey carried out among professional philosophers, 59 percent identified as compatibilists.

The second biggest camp among philosophers are libertarians, who argue that free will is incompatible with determinism, but that, because free will exists, determinism must be false. I will not discuss libertarianism because, well, it's incompatible with what we know about nature.

Let us therefore talk a little more about compatibilism, the philosophy that Immanuel Kant charmingly characterized as a "wretched subterfuge," that the nineteenth-century philosopher William James put down as a "quagmire of evasion," and that the contemporary philosopher Wallace Matson called out as "the most flabbergasting instance of the fallacy of changing the subject." Yes, indeed, how do you get free will to be compatible with the laws of nature, keeping in mind that—let's not forget—the future is fixed except for occasional quantum events that we cannot influence?

One thing you can do to help your argument is to improve physics a little bit. The philosopher John Martin Fischer has dubbed philosophers who do this "multiple-past compatibilists" and "local-miracle compatibilists." The former argue that your actions change the past to something that it wasn't. The latter argue that supernatural events beyond the laws of nature allow your decision to somehow avoid the predictions of theories that have been confirmed countless times. I will not discuss these here either, because this is a book about what we can learn from physics, not about how we can creatively ignore physics.

Among the compatibilist ideas that are at least not wrong, the most popular one is that your will is free because it's not predictable, certainly not in practice and possibly not even in principle. This position is maybe most prominently represented by Daniel Dennett. If you want to think about free will that way, fine. But the future is still fixed except for occasional quantum events that we cannot influence.

The philosopher Jenann Ismael has furthermore argued that free will is a property of autonomous systems. By this she means that different subsystems of the universe differ in how much their behavior depends on external input versus internal calculation. A toaster, for example, has very little autonomy—you push a button and it reacts. Humans have a lot of autonomy because their deliberations can proceed mostly decoupled from external input. If you want to call that free will, fine. But the future is still fixed except for occasional quantum events that we cannot influence.

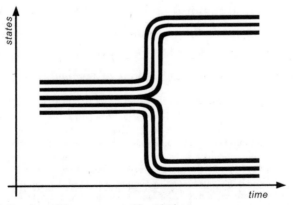

Figure 11: How free will becomes compatible with physics.

There are quite a few physicists who have backed compatibilism by finding niches in which to embed free will into the laws of nature.[a] Sean Carroll and Carlo Rovelli suggest that we should interpret free will as an emergent property of a system. A pepped-up version of this argument was recently put forward by Philip Ball. It relies on using causal relations between macroscopic concepts—so also emergent properties—to define *free will*.

Remember that emergent properties are those that arise in approximate descriptions on large scales when details of the microphysics have been averaged. Figure 11 illustrates how this can be used to make place for free will. On the microscopic level, the paths (white lines) are determined by the initial value—that is, the place they start from on the left. But on the macroscopic level, if you forget about the exact initial conditions and look at the collection of all microscopic paths, the macroscopic path (black outline) forks.

The above-mentioned physicists now say that if you ignore the determined behavior of particles on the microscopic level, then you can no longer make predictions on the macroscopic level. The paths fork: hurray! Of course, this happens just because you've ignored what's really going on. Yeah, you can do that. But the future is still fixed except for occasional quantum events that we cannot influence. When Sean Carroll summed up his compatibilist stance with "free will is as real as a baseball," he should have added "and equally free."

Having said that, I don't have a big problem with physicists' or philosophers' compatibilist definitions of *free will*. After all, they're

[a]The physicists (myself included) don't usually discuss whether free will is compatible with determinism (which is the classical libertarianism/compatibilism divide) but whether it's compatible with the laws of nature, taking into account that quantum mechanics (in the standard interpretation) has a fundamentally random element. This distinction doesn't make a difference, because there is no "will" in quantum randomness, but it sometimes leads to confusion. For example, a compatibilist physicist might well answer "no" when asked whether human actions are determined by the initial state of the universe or can be predicted from perfect information, yet that answer would, according to some surveys, put them in the libertarian camp.

just definitions, neither right nor wrong, merely more or less useful. But I don't think such verbal acrobatics address the issue that normal people—sorry, I mean nonphilosophers—worry about. A 2019 survey of more than five thousand participants from twenty-one countries found that "across cultures, participants exhibiting greater cognitive reflection were more likely to view free will as incompatible with causal determinism." It seems we weren't born to be compatibilists. That's why, for many of us, learning physics shakes up our belief in what we think of as free will—as it did for me. This, in my mind, is the issue that needs to be addressed.

○ ○ ○

As you see, it isn't easy to make sense of free will while respecting the laws of nature. Fundamentally, the problem is that, for all we currently know, strong emergence isn't possible. That means all higher-level properties of a system—those on large scales—derive from the lower levels where we use particle physics. Hence, it doesn't matter just how you define *free will*; it'll still derive from the microscopic behavior of particles—because everything does.

The only way I can see to make sense of free will is therefore that the derivation from the microscopic theory fails for some cases for some reason. Then strong emergence could be an actual property of nature and we could have macroscopic phenomena—free will among them—that are truly independent of the microphysics. We don't now have a shred of evidence that this indeed happens, but it's interesting to think about what it would take.

To begin with, the mathematical techniques we use to solve the equations that relate microscopic with macroscopic laws don't always work. They often rely on certain approximations, and when those approximations aren't adequate to describe the system of interest, we just don't know what to do with the equations. This is a practical problem, for sure, but it doesn't matter insofar as the properties of the laws are concerned. The relation between the deeper and higher levels doesn't go away just because we don't know how to solve the equations that relate them.

What gets us a little closer to strong emergence are two examples where physicists have studied the question whether composite systems can have properties whose value is undecidable for a computer. If that were so, that would be a much better argument for a macroscopic phenomenon that is "free" of the microphysics than just saying we don't know how to compute it. It'd actually prove that it can't be computed. But these two examples require infinitely large systems for the argument to work. The statement then comes down to saying that for an infinitely large system, certain properties cannot be calculated on a classical computer in finite time. It's not a situation we'd encounter in reality and, hence, it doesn't help with free will.

However, it could be that the derivation of the macroscopic behavior from the microphysics fails for another reason. It could be that in the calculation we run into a singularity beyond which we just cannot continue—neither in practice nor in principle. This does not necessarily bring back infinites again, for in mathematics a singular point is not always associated with something becoming infinite; it's just a point where a function can't be continued.

We currently have no reason to think this happens for the actual microphysics that is realized in our universe, but it's something that could conceivably turn out to be so when we understand the math better. So if you want to believe in a free will that's truly governed by natural laws independent from those of elementary particles, the possibility that the derivation of the macrophysical laws runs into a singular point seems to me the most reasonable one. It's a long shot, but it's compatible with all we currently know.

Life without Free Will

The American science writer John Horgan calls me a "free will de-
nier," and by now you probably understand why. But I certainly do not
deny that many humans have the impression they have free will. We
also, however, have the impression that the present moment is special, which we already saw is an illusion, and if I went by my impression, I'd say the horizontal lines in figure 12 aren't parallel. If my
research in the foundations of physics has taught me one thing, it's
that one shouldn't count on personal impressions. It takes more than
an impression to infer how nature really works.

Despite the limitations of our brains, not to brag, but I think we
humans have done a pretty good job figuring out the laws of nature.
We *did*, after all, come to understand that "now" is an illusion, and,
using your brain at its finest, you can go and measure the lines in figure 12 to convince yourself they really are parallel. They still won't
look parallel, but you will know they're parallel nevertheless. I think
we should deal with free will the same way: leave aside our intuitive
feelings and instead follow reason to its conclusions. You will still *feel*
as if you have free will, but you will know that really you're running
a sophisticated computation on your neural processor.

Figure 12: Café wall illusion. The horizontal lines are parallel.

But I'm not trying to missionize you. As I said, it all depends on how you define *free will*. If you prefer a compatibilist definition of *free will* and want to continue using the term, science says nothing against that. Therefore, to answer this chapter's question: Has physics ruled out free will? No, it has merely ruled out certain ideas about free will. Because for all we currently know the future is fixed except for occasional quantum events that we cannot influence.

How can we deal with this? I get asked this a lot. It seems to me the problem is that many of us grow up with intuitive ideas about how our own decision-making works, and when these naive ideas run into conflict with what we learn about physics, we have to readjust our self-image. This isn't all that easy. But there are a few ways to sort it out.

The easiest way to deal with it is through dualism, according to which the mind has a nonphysical component. Using dualism, you can treat free will as an ascientific concept, a property of your soul, if you wish. This will be compatible with physics as long as the non-physical component does not interact with the physical one, because then it'd be in conflict with evidence—it'd become physical. Because the physical part of our brain is demonstrably the thing we use to make decisions, I don't see what one gains from believing in a non-physical free will, but then this isn't a new problem with dualism, and at least it isn't wrong.

You are also welcome to use the little loophole in the derivation from microphysics that I detailed earlier. Though I suspect if you tell someone you think free will is real because the renormalization group equations might run into an essential singularity, you might as well paint GEEK on your forehead.

Personally, I think the best way to deal with the impossibility of changing the future is to shift the way we think about our role in the history of the universe. Free will or not, we are here, and therefore we matter. But whether ours will be a happy story or a sad story, whether our civilization will flourish or wither, whether we will be remembered or forgotten—we don't yet know. Instead of thinking of ourselves as selecting possible futures, I suggest we remain curious about what's to come and strive to learn more about ourselves and the universe we inhabit.

I have found that abandoning the idea of free will has changed the way I think about my own thinking. I have begun paying more attention to what we know about the shortcomings of human cognition, logical fallacies, and biases. Realizing that in the end I am just working away on the input I collect, I have become more selective and careful with what I read and listen to.

Odd as this may sound, in some circumstances I've had to work hard to convince myself to listen to myself. For example, I commuted for several years between Germany and Sweden, racking up dozens of flights a year. Yet for some reason it didn't occur to me to sign up for a frequent flyer card. When someone asked me about it two years into my commute, I felt rather dumb. But instead of immediately signing up for a frequent flyer card, I put it off, reasoning that I'd forgone so many benefits already, I might as well not bother. It's a curious instance of loss aversion ("throwing good money after bad"), though in this case it wasn't an actual loss but an absence of benefits. Recognizing this, I did eventually sign up for a card. If I hadn't known it was a cognitive bias, I don't think I would have; I'd instead have done my best to forget about it altogether, thereby working against my own interests.

I am not telling you this because I'm proud I made a rational decision (in this instance, at least). To the very contrary, I am telling you this to highlight that I'm as irrational as everybody else. Yet I think I have benefited from accepting that my brain is a machine—a sophisticated machine, all right, but still prone to error—and it helps to know which tasks it struggles with.

When I explain that I don't believe in free will, most people will joke that I couldn't have done otherwise. If you had this joke on your mind, it's worth contemplating why it was easy to predict.

Free Will and Morals

On January 13, 2021, Lisa Marie Montgomery became the fourth woman to be executed in the United States, the first in sixty-seven years. She was sentenced to death for the murder of Bobbie Jo Stinnett, age twenty-three and, at the time of the crime, eight months pregnant. In 2004, Montgomery befriended the younger woman. On December 16 of that year, she visited Stinnett and strangled her. Then she cut the unborn child from the pregnant woman's womb. For a few days, Montgomery pretended the child was hers, but she confessed quickly when charged by police. The newborn remained unharmed and was returned to the father.

Why would someone commit a crime as cruel and pointless as this? A look at Montgomery's life is eye-opening.

According to her lawyers, Montgomery was physically abused by her mother from childhood on. Beginning at age thirteen, she was regularly raped by her stepfather—an alcoholic—and his friends. She repeatedly, but unsuccessfully, sought help from authorities. Montgomery married young, at the age of eighteen. Her first husband, with whom she had four children, also physically assaulted her. By the time of the crime, she had been sterilized, but sometimes pretended to be pregnant again. Once in prison, Montgomery was diagnosed with a whole list of mental health problems: "bipolar disorder, temporal lobe epilepsy, complex post-traumatic stress disorder, dissociative disorder, psychosis, traumatic brain injury and most likely fetal alcohol syndrome."

I would be surprised if the previous paragraph did not change your mind about Montgomery. Or, if you knew her story already, I would be surprised if you did not have a similar reaction the first time you heard it. The abuse she suffered at the hands of others doubtless contributed to the crime. It left a mark on Montgomery's psyche and personality that contributed to her actions. To what extent was she even responsible? Wasn't she herself a victim, failed by institutions that were supposed to help her, too deranged to be held accountable? Did she act out of her own free will?

We frequently associate free will with moral responsibility in this way, which is how it enters our discussions about politics, religion, crime, and punishment. Many of us also use free will as a reasoning device to evaluate personal questions of guilt, remorse, and blame. In fact, much of the debate about free will in the philosophical literature concerns not whether it exists in the first place but how it connects to moral responsibility. The worry is that if free will goes out the window, society will fall apart because blaming the laws of nature is pointless.

I find this worry silly. If free will doesn't exist, it has never existed, so if moral responsibility has worked so far, why should it suddenly stop working just because we now understand physics better? It's not as though thunderstorms changed once we understood they're not Zeus throwing lightning bolts.

The philosophical discourse about moral responsibility therefore seems superfluous to me. It is easy enough to explain why we—as individuals and as societies—assign responsibility to people rather than to the laws of nature. We look for the best strategy to optimize our well-being. And trying to change the laws of nature is a bad strategy.

Again, one can debate just exactly what *well-being* means; the fact that we don't all agree on what it means is a major source of conflict. But just exactly what it is that our brains try to optimize, and what the difference is between your and my optimization, isn't the point here. The point is that you don't need to believe in free will to argue that locking away murderers benefits people who could potentially be murdered, whereas attempting to change the initial conditions of the universe doesn't benefit anyone. That's what it comes down to: We evaluate which actions are most likely to improve our lives in the future. And when it comes to that question, who cares whether philosophers have yet found a good way to define *responsibility*? If you are a problem, other people will take steps to solve that problem—they will "make you responsible" just because you embody a threat.

That way, we can rephrase any discussion about free will or moral responsibility without using those terms. For example, instead of questioning someone's free will, we can debate whether jail is really the most useful intervention. It may not always be. In some circumstances, mental health care and preventions against family violence may be the longer lever to reduce crime. And of course, there are other factors to consider, such as retaliation and deterrence and so on. This isn't the place to have this discussion. I merely mean to demonstrate that it can be had without referring to free will.

The same can be done for personal situations. Whenever you ask a question like "But could they have done differently?" you are evaluating the likelihood that it may happen again. If you come to the conclusion that it's unlikely to happen again (in terms of free will you might say, "They had no other choice"), you may forgive them ("They were not responsible"). If you think it's likely to happen again ("They did it on purpose"), you may avoid them in the future ("They were responsible"). But you can rephrase this discussion about moral responsibility in terms of an evaluation about your best strategy. You could, for example, reason: "They were late because of a flat tire. It's therefore unlikely to happen again, and if I was angry about it, I might lose good friends." Free will is utterly unnecessary for this.

Let me be clear that I don't want to tell you to stop referring to free will. If you find it's handy, by all means, continue. I just wanted to offer examples for how one can make moral judgments without it. This matters to me because I am somewhat offended by being cast as morally defunct just because I agree with Nietzsche that free will is an oxymoron.

The situation is not helped by the recurring claim that people who do not believe in free will are likelier to cheat or harm others. This view was expressed, for example, by Azim Shariff and Kathleen Vohs in a 2014 *Scientific American* article in which they said their research shows that "the more people doubt free will, the more lenient they become toward those accused of crimes and the more willing they are to break the rules themselves and harm others to get what they want."

First, let us note that—as is often the case in psychology—other studies have given different results. For example, a 2017 study on free will and moral behavior concluded, "We observed that disbelief in free will had a positive impact on the morality of decisions toward others." The question is the subject of ongoing research; the brief summary is that it's still unclear how belief in free will relates to moral behavior.

More insightful is to look at how these studies were conducted in the first place. They usually work with two separate groups, one primed to doubt free will, the other one a neutral control group. For the no-free-will priming, it has become common to use passages from Francis Crick's 1994 book, *The Astonishing Hypothesis: The Scientific Search for the Soul*. Here is an excerpt:

> You, your joys and your sorrows, your memories and your ambitions, your sense of personal identity and free will, are in fact no more than the behavior of a vast assembly of nerve cells and their associated molecules. Who you are is nothing but a pack of neurons.

This passage, however, does more than just neutrally inform people that the laws of nature are incompatible with free will. It also denigrates their sense of purpose and agency by using phrases like "no more than" and "nothing but." And it fails to remind the reader that this "vast assembly of nerve cells" can do some truly amazing things, like reading passages about itself, not to mention having found out what it's assembled of to begin with.

Of course, Crick's passage is deliberately sharply phrased to get his message across (not unlike some passages you read in chapter 4 of this very book); nothing wrong with that in and of itself. But it doesn't prime people just to question free will; it primes them for fatalism—the idea that it doesn't matter what you do. Suppose he'd primed them instead using this passage:

> You, your joys and your sorrows, your memories and your ambitions, your sense of personal identity and free will, are the result of a delicately interwoven assembly of nerve cells and their associated molecules. That pack of neurons is the product of billions of years of evolution. It endows you with an unparalleled ability to communicate and collaborate, and a capacity for rational thought superior to that of all other species.

Not as punchy as Crick's version, I admit, but I hope it illustrates what I mean. This version also informs readers that their thoughts and actions are entirely a result of neural activity. It does so, however, by emphasizing how remarkable our thinking abilities are. It would be interesting to see whether people primed in this way to disbelieve in free will are still more likely to cheat on tests, don't you think?

>> THE BRIEF ANSWER

According to the currently established laws of nature, the future is determined by the past, except for occasional quantum events that we cannot influence. Whether you take that to mean that free will does not exist depends on your definition of *free will*.

IS CONSCIOUSNESS COMPUTABLE?

An Interview with Roger Penrose

When I enter Roger Penrose's office, he is bent over his desk, nose three inches from a laptop screen. He squints through thick glasses at slides for a talk he is giving later. Now in my mid-forties, I think of myself as old, but realize Roger is more than twice my age. He has been collecting awards, medals, and honors since before I was born. Now Roger, emeritus professor of mathematics at the University of Oxford, has so many things named after him— Penrose tilings, Penrose diagrams, the Penrose triangle, the Penrose process, the Penrose-Hawking theorems—it's hard to believe he hasn't won a Nobel Prize. Little do I know, in 2019, that he will go on to win the prize a year later.

He briefly glances at me and vaguely apologizes for the delay while he sets the font size to what looks like 200 points. I assure him it's not an issue and unpack my notepad and sound recorder.

I am here, at the University of Oxford, to attend a conference about mathematical models of consciousness, but Roger kindly agreed to an interview during one of the conference breaks. His talk is due after the break, and he is busy rearranging his slides, an eclectic mix of mathematics, quantum mechanics, cosmology, and neurobiology.

Besides applying his mathematical skills to solve physics problems—such as the question whether stars can collapse to black holes—Roger has put forward numerous speculative ideas for how nature fundamentally works, among them that gravity causes the reduction of the wave function and that the universe is cyclic, going from expansion to contraction and into a new Big Bang infinitely often. There would be many things to talk about, but I am particularly interested in his views on consciousness.

Again I begin by asking, "Are you religious?"

"Not religious in any sense of the term that people usually mean it," he says.

"In any other form?"

"Do I believe in a god?" Roger asks himself. "No, not in the usual sense of the word."

"Do you believe that the universe has a purpose?" I guess, sensing he wants to add something.

"You're getting close . . . ," Roger says hesitantly. "I don't know if the universe has a purpose, but I would say that there is something more to it, in the sense that the presence of conscious beings is probably something deeper, not just not random. It's hard to say. It's not that I have anything clear to say about what I believe. I just don't think saying it's chance is a sufficient explanation."

"Do you think that consciousness fits into the framework that physicists have come up with so far?" I inquire.

"No," Roger says. "This is a belief I have had for a long time. When I was an undergraduate, I was very troubled by what I thought I heard about the Gödel theorem, which seemed to say that there were things in mathematics that we could not prove. And then I went to this course given by [mathematician] Stourton Steen. The way he described Gödel's theorem was not that there were things you couldn't prove. He explained that you could have a logical system which could in principle be put on a computer, and if you feed in a mathematical theorem, it chugs away and tells you whether the theorem is true or false or goes on forever. This system is meant to be following reasoning that you trust; otherwise, what's the point? So it follows the rules and if it says a theorem is true, then you believe it is true.

"But following these rules, you can construct another mathematical statement, that is the Gödel statement. And you can see from the way that it's constructed that the Gödel statement is true. Its truth is derived from your belief that the system only gives you truth. Yet you can show by the construction of the Gödel statement that the computer cannot derive [that] it is true."

Gödel's famous (second) incompleteness theorem is usually presented as a statement about sets of mathematical axioms—that is, assumptions from which you draw logical conclusions. Gödel showed that the consistency of any set of axioms (that's at least as complex as the set of the natural numbers) is unprovable. He formulated statements—*Gödel statements* or *Gödel sentences*—that are true, but one cannot *prove* them to be true within the system of axioms. Penrose interprets this to mean that we humans are able to recognize a truth that a computer algorithm, fed only the axioms whose consistency is in question, cannot see. If the algorithm could see the truth, it could prove it, contradicting Gödel's theorem. I have some more to tell you about this in chapter 9, but let's first hear what else Roger has to say.

"I find that stunning, because it's telling us that your belief that the system works is stronger than the system itself. What are you doing that enables you to transcend the system? What's going on there? To me that's a very clear illustration of the power of understanding. I don't know what understanding is, but it seems to me that it can't be computation. Whatever is going on in conscious understanding is not the same as a complicated computation."

This doesn't really make sense to me. I ask, "How do you square this with knowing that we are all ultimately made of particles and these particles follow computable equations?"

"Yes. How does this work?" Roger says and nods. "At first I thought to myself, 'Maybe it's the continuum. That is why it's not, strictly speaking, a computation.' But I don't think that's it. You can put Newtonian mechanics and general relativity on a computer, and you can make that calculation as accurate as you like. Then I thought, 'What about quantum mechanics?' There is the Schrödinger equation, but that's still a computation. But then there is the measurement process. I thought, 'Well, that's a big gap in our understanding,' and I think there has to be some theory about what is really going on in this reduction of the quantum state. And since that was the only gap I could see, I thought this had to be it."

He laughs and continues. "I did have the idea that when I retired—that seemed like a long time in the future then, but is now a long time in the past—when I retired, I would write a book. That was *The Emperor's New Mind*, which I actually wrote before I retired. First of all, I would explain what I knew about physics, and then I would try to learn something about neurophysiology and synapses and the funny way they operate. I thought that by the time I'd learned all that, I would see a place where quantum-state reduction could play a role. But I didn't. So the book faded out in the end, in my view. I gave a rather silly idea, which I didn't believe in, and then I stopped the book.

"You see," Roger explains, "I was hoping my book might stimulate young people to look at it. But the only people I heard from were retired people. Apparently they were the only people who had the time to read the book! And then I got a letter from Stuart Hameroff that said, 'You need to look at these little tubes, the microtubules.' I get a lot of crazy crackpot letters, and I thought, 'Here's another one.' But then I looked it up, and I thought I ought to have known of this before. This is a good place for quantum coherence."

Roger Penrose and Stuart Hameroff would go on to write several papers together about microtubules, which are small tubes of proteins that can be found in cells, including neurons. The idea is that collections of microtubules in neurons can display quantum behavior. When these quantum states of the microtubules are reduced—that is, the quantum effects disappear—conscious awareness arises and free will becomes possible. This conjecture is called *orchestrated objective reduction*, or *Orch OR* for short.

Orchestrated objective reduction has been met with skepticism by scientists, in physics as well as in neurobiology. The major reason is that in standard quantum mechanics, microtubules couldn't maintain quantum effects remotely long enough to play any role in neural activity. This means it'd take a significant modification of quantum mechanics to make the idea work. Indeed, this is what Penrose and Hameroff argue happens. Possible, but far-fetched and lacking evidence. It has also remained unclear what the loss of quantum effects in microtubules has to do with consciousness or free will.

As you can tell, I am not convinced by the microtubules business. Nevertheless, I am intrigued by the link Roger sees between consciousness and wave function reduction.

"If I may summarize this," I suggest, "you are saying that the quantum measurement process is the gap that we have in the foundations of physics, and if understanding is not a computation, then this is where it has to come in. So the measurement process depends on human consciousness?"

"You have to read it the right way around," Roger says. "Many people in the foundations of quantum mechanics, including John von

Neumann and Eugene Wigner, had the view that somehow the reduction of the state was caused when a conscious being looked at it. This didn't make much sense to me."

He offers an example: "Imagine a space probe going out looking at planets. The space probe visits a planet with no conscious being anywhere, not on that planet and not anywhere close, and it takes a photograph. Now, the weather is a chaotic system and ultimately depends on quantum effects. So the space probe sees a superposition of different kinds of weathers. It takes a photo and sends it back to Earth. After I-don't-know-how-many years, someone sees the photo on a screen. And when that conscious being sees the photo, flip, suddenly it becomes one weather? That makes absolutely no sense to me. It seems to me that is not the right answer, surely."

"So it's not that consciousness causes the reduction of the wave function but that the reduction of the wave function plays a role in consciousness?"

"Yes," Roger says. "And that's not how people thought about it. I am quite surprised that so few people thought about it this way around. The idea is that there is something going on in brain processing. Microtubules probably play a role, but it may not be the only thing. The question is what is it that induces the state to collapse. It's got to be something fundamental, and it's got to be something outside standard quantum mechanics."

>> THE BRIEF ANSWER

If consciousness emerges from the fundamental laws of physics that we already know, it is computable. However, the update of the wave function in quantum mechanics might signal that we are missing some part of the story, and this missing part might be uncomputable.

If that is so, consciousness might also be uncomputable. This wouldn't mean that consciousness causes the update of the wave function; rather, it is the other way round: that the update of the wave function would play a role in conscious awareness. This is highly speculative and no evidence speaks for it, but at present it's compatible with what we know.

Chapter 7

WAS THE UNIVERSE
MADE FOR US?

Imagine No Religions

At birth, we can neither walk nor focus our gaze nor do as much as ask a question. As we grow, our world expands. We explore the crib, the room, the apartment and its balcony. We go on a first trip to the playground. There's school, college, the first time on a plane. We realize that we live on a planet with more than seven billion people and counting, and that the particular culture we have grown up in is only one of hundreds. Planet Earth, we learn, is billions of years old, modern civilization is just a blip on the time line, and the dots on the night sky are other stars, some of them entire galaxies, in a universe that might well be infinitely large.

Our exploration of the world comes with the recognition of our own insignificance, and science has made this message only starker. The universe is big and we are small, merely some creatures crawling around on a medium-size rocky planet, one of an estimated 100 billion planets in one of about 200 billion galaxies in the visible universe. We quite literally don't matter: most of the matter in the universe—about 85 percent—is dark matter, not the stuff we are made of, and in any case, whatever we achieve, it'll be wiped out by entropy increase eventually.

Some find comfort in this insignificance; others find it disturbing. They'd rather humans played a preferred role. Certainly our own existence must mean something, they insist. Isn't it peculiar, they ask, that the universe is the way it is, so we can be the way we are? Isn't there something special about it?

The question whether our universe is especially well suited for the development of life, whether our existence signals the presence of an intelligent being who set up things "just right," hovers at the border between science and religion. The position that the universe requires a creator has been taken, for example, by the philosopher and theologian Richard Swinburne but also by the astrophysicists Geraint Lewis and Luke Barnes, who argue that their views are based on science. The very opposite opinion has been put forward most prominently by Stephen Hawking, according to whom we live in a multiverse that eliminates the need for a creator.

These arguments sound like the exact opposite of each other: one claims a creator is necessary; the other claims it's unnecessary. Nevertheless, they are similar in that they are both ascientific. They both postulate the existence of things that are unnecessary to describe what we observe.

o o o

The issue is this. The currently known laws of nature contain twenty-six constants. We can't calculate those constants; we just determine their values by measurement. The *fine-structure constant* (α) sets the strength of the electromagnetic force. *Planck's constant* ($\hbar$) tells us when quantum mechanics becomes relevant. *Newton's constant* (G) quantifies the strength of gravity. The *cosmological constant* (Λ) determines the expansion rate of the universe. Then there are the masses of the elementary particles, and so on.

Now you can ask, "What would a universe look like in which one or several of these constants had a value a little different from those we measure?" Imagine God sitting in front of a big panel with knobs. Each knob has the name of a constant on it. With a mischievous grin, God turns some of the knobs away from the values they have in our universe. Suddenly, humans disappear.

Change too many of the constants of nature, and processes that are essential for life as we know it could not happen, and we could not exist. For example, if the cosmological constant was too large, then galaxies would never form. If the electromagnetic force was too strong, nuclear fusion could not light up stars. There's a long list of calculations of this type, but they're not the relevant part of the argument, so I don't want to go through them.

Let me instead cut right to the relevant part of the argument, which goes like this: It's extremely unlikely that these constants would just coincidentally happen to have exactly the values that allow for our existence. Therefore, the universe as we observe it requires an explanation, a god who fine-tuned the knobs. If not a god, then we need another explanation. The multiverse hypothesis allegedly is one. If there is a universe for any possible combination of constants, the argument goes, then certainly ours must be among them, so that explains it.

However, the multiverse hypothesis doesn't explain anything. A good scientific hypothesis is one that is useful for calculating the outcomes of measurements. You can therefore tell whether a hypothesis is any good just by looking at whether scientists actually—and successfully—use it to calculate measurement outcomes. But no one uses the multiverse hypothesis to calculate anything of practical interest. That's because to calculate the outcome of observations in our universe, you need the values of these constants. Merely declaring that "they exist" doesn't help.

When physicists nevertheless try to do calculations using the multiverse, the results (or rather, absence of results) can be hilarious. For these calculations, physicists assume the different types of universes (different values of the constants) have certain probabilities to exist. This is called a *probability distribution*. The probability distribution for a fair die, for example, is ⅙ for each side.

The probabilities for other universes to exist aren't measurable, because we can't measure what we can't observe, so physicists will just postulate something. If they then try to calculate the probability for some observation in our universe, that merely rephrases whatever they postulated, so one doesn't learn anything from it—garbage in, garbage out. But it creates a new problem, namely that now they have to explain what the probability is that someone observes something in the multiverse in the first place. And what does "someone" mean in a universe with different laws of nature?

A few years ago, for example, a group of astrophysicists tried to use the multiverse hypothesis to find out how likely it is that galaxies look the way they look and that the cosmological constant is what it is. For this, they used computer simulations to see how galaxies form in universes with different cosmological constants. Here is an extract from the paper:

> We might wonder whether any complex life form counts as an observer (an ant?), or whether we need to see evidence of communication (a dolphin?), or active observation of the universe at large (an astronomer?). .

We already knew, of course, that not all values of the cosmological constant are compatible with our observations, because this constant determines how fast the universe expands, and if it expands too fast, galaxies are ripped apart. It is certainly nice to see how this happens in a computer simulation, but elaborations about dolphins in the multiverse don't bring further insights; they just add an arbitrary, unobservable probability distribution over unobservable universes. The authors elaborate on their conundrum:

> What would it mean to apply two different [probability distributions] to this model, to derive two different predictions? How could all the physical facts be the same, and yet the predictions of the model be different in the two cases? What is the [probability distribution] about, if not the universe? Is it just our own subjective opinion? In that case, you can save yourself all the bother of calculating probabilities by having an opinion about your multiverse model directly.

Indeed, you can save yourself the bother. That must be the most honest discussion in the scientific literature, ever.

So if the multiverse doesn't explain the values of the constants, then does this mean we need a creator? No, that conclusion is equally ascientific, because from the scientific perspective there is nothing in need of an explanation. The fine-tuning argument for a creator rests on claiming that the values we observe for the constants are unlikely. But there is no way to ever quantify this probability, because we will never measure a constant of nature that has a value other than the one it does have.

To quantify a probability, you have to collect a sample of data. You could do that, for example, if you were throwing dice. Throw them often enough, and you get an empirically supported probability distribution. But we do not have an empirically supported probability distribution for the constants of nature. And why is that? It's because (drums, please) they are constant.[a] Saying that the only value we have ever observed is "unlikely" is a scientifically meaningless statement. We have no data, and will never have data, that allow us to quantify the probability of something we cannot observe. There's nothing quantifiably unlikely, and therefore there's nothing in need of explanation.

An example. If you blindly reach into a bag and pull out a sheet of paper with the number 77974806905273, do you exclaim, "Wow, how amazingly unlikely! That requires an explanation"? Probably not, because you have no idea what else is in the bag. For all you know it could still contain a trillion papers with the same number, your missing right sock, turtles all the way down, all of that together, or nothing at all. If you pull out only one number, you know nothing about the probability of getting that number. It's the same for the constants of nature. We have pulled one set of numbers, but that was our only pull. We have no idea whether that was likely or unlikely—and we'll never know.

You can, of course, just assume a probability distribution for the constants of nature in order to make the fine-tuning argument work, just as you do in the multiverse. But that creates the same problem. The conclusions about how likely or unlikely our universe is will just give back what you put in. There are probability distributions according to which the constants we measure are unlikely, and others according to which they are likely. It's just that the people who argue that our universe is fine-tuned don't use the latter, because doing so wouldn't lead to the conclusion they want.

Simply put, claiming that the constants of nature are fine-tuned

---

[a]Some physicists have proposed theories in which the constants of nature are replaced with parameters that can change with time or place, but that's a different story entirely and has nothing to do with the fine-tuning arguments.

for life is not a scientifically sound argument, because it depends on arbitrary assumptions. While a creator or a multiverse is not ruled out by science, science does not require their existence either.

o o o

I recently took part in a debate about the question whether the universe is fine-tuned, organized by a British Christian institution. My discussion partner was Luke Barnes, who argues that the values of the constants of nature require an explanation; he is one of the authors of the 2016 book *A Fortunate Universe: Life in a Finely Tuned Cosmos*.

I didn't look forward to the debate. I have found it futile to argue with fine-tuning believers. They just aren't interested in separating the scientific from the ascientific part of their argument. Also, I am terribly unspontaneous. If you put me on the spot, I can't find answers to the most obvious questions. Hell, I'll sometimes mispronounce my own name. Full disclosure: the major reason I agreed to this debate is that they paid for it.

At the time of the debate, early 2021, both the UK and Germany were in lockdown due to the COVID pandemic, so the event took place online. I called in from Germany, Barnes from Australia, and our host remained in the UK.

Barnes turned out to be a big-faced, middle-aged man with full hair and a lot of beard. He'd positioned himself in front of a bookshelf with his own books on display. Speaking to him, I realized immediately that he is a first-rate astrophysicist; he understands his material in depth, both the observation and the theory. And he did what many physicists do in response to my fine-tuning criticism: pointed out that I am using the *frequentist interpretation* of probability, not the

Bayesian interpretation. That's correct—but I do this because other-wise the fine-tuning claim can't even be formulated.

You see, in the frequentist interpretation, probabilities quantify relative numbers of occurrences. It's the interpretation usually taught in school, so you are probably familiar with it. Frequentist probabilities are objective; they are statements about what happens. In the Bayesian interpretation, probabilities are instead statements about your expectation given your prior belief (usually just called the *prior*). These probabilities are by construction subjective.

Using the Bayesian interpretation, therefore, the fine-tuning argument comes down to saying, "Based on my prior belief that the constants of nature could have been anything, I am surprised they are what they are." But this doesn't mean the universe is fine-tuned for life; it just means you expected something that turned out to not be the case. Big deal. The statement "Based on my prior belief that I could have woken up being anything, I am surprised I'm human" likewise doesn't mean it was ever likely you'd wake up being a verminous monster. More likely, you've read too much Kafka.

In our debate, Luke Barnes readily agreed it's not a scientific argument to claim that the constants of nature require an explanation. Based on my prior belief that scientists tend to be unwilling to admit to ascientific arguments, I was surprised.

Thomas Bayes, by the way, after whom Bayesian probabilities are named, was a Presbyterian minister in eighteenth-century England. Fittingly enough, the first known application of Bayes's probability calculus was the attempt to prove that God exists. The proof didn't convince anyone who wasn't already convinced, but some ideas, it seems, don't go out of fashion.

Do We Live in the Best of All Possible Worlds?

When I was introduced to physics in middle school, I didn't like it. It was a stream of equations relating variables whose meaning I constantly forgot, and the only purpose of the discipline seemed to be to massage said equations into new forms. Wasn't there some minimal set of equations, I wanted to know, from which all the rest could be derived? And if that was so, why were we taught all the clutter?

In reply I was told that a theory of everything was almost certainly a pipe dream. Even Einstein failed to find one, and the clutter was here to stay, at least for now—and here's this week's homework.

It wasn't really a theory of everything that I'd been after; I had just hoped we could wrap up some years of physics in a month and get it over with. But now that you've said it, I thought, a theory of everything sounds like a good idea.

The clutter stayed throughout my school education. But much of it suddenly vanished in the first semester of university physics, when the *principle of least action* was introduced. It was a revelation: there was indeed a procedure to arrive at all those equations! Why hadn't anybody told me?

Today I think they don't teach the principle of least action in school because then everybody would go and study physics. So, with the warning that you might get hooked, here's how it works.

For every system you want to describe (say, a swinging pendulum), there's a function called the *action* (usually denoted S), which takes on the smallest possible value for that behavior of the system that is actually realized in nature. That is, if you consider all possible things a system could do and you calculate the action for each of them, the case you actually observe is the one with the minimal action. This doesn't mean that the system (the pendulum) actually tries all possible motions; it's just that the motion you observe is that with the least action.

The principle of least action was anticipated in the seventeenth century by Pierre de Fermat, who found that the path taken by a ray of light through a medium is that which requires the least amount of time. But it's really a much more general principle. The requirement that the action take on a minimal value leads to an evolution law. You choose initial conditions, and then you "only" have to solve the equations.

The *action* here has nothing to do with the *action* in *action movies*. It's there merely to quantify Gottfried Wilhelm Leibniz's idea that we live in the "best of all possible worlds." You just have to tell God that *best* means minimizing the action. But what is this mysterious quantity, the action?

In the first semester of physics, there's an action for the pendulum, one for throwing a stone, one for the orbits of the planets—you get the idea. So you have a recipe for calculating what a system does, but there are still all those different actions.

These actions, however, are different not because the physics is different but because the systems are different. They might have different arrangements, or you might be describing them on different levels of resolution. Remember, we have all these effective theories.

If you're throwing a stone, for example, you usually assume the gravitational field is constant in a vertical direction. That's a good approximation but, strictly speaking, not correct. A better approximation is that the Earth's gravitational field is spherically symmetrical and falls with the inverse square of the distance to Earth's center. An even better approximation is to use the exact distribution of matter that makes up our planet and calculate the gravitational field from it.

So instead of using an action that assumes what the gravitational field is, you could add to the action a term that is minimal for the gravitational field, and then the principle of least action allows you to calculate both the motion of the stone and the gravitational field. And if you do that, then planetary orbits and throwing a stone become pretty much the same thing, except for the initial condition by which you specify what matter is located where and what its initial velocity is.

That's as long as you neglect air friction, which the stone has but a planet that goes around the Sun doesn't. For the stone, you would therefore also have to take into account the interaction of the stone's molecules with the air molecules, and the air molecules' interaction with themselves. And then you would begin to note what we already saw earlier, that once you get down to interactions on atomic scales, you can no longer ignore quantum mechanics.

In quantum mechanics, the principle of least action works a little differently. According to the path-integral approach pioneered by Richard Feynman, a quantum-mechanical system takes not only the path of least action but all possible paths. Each path makes a contribution to what's called the *amplitude* of the system, and the absolute square of the amplitude gives you the probability for the system to end up at a certain endpoint.

Because the paths' contributions to the amplitude don't necessarily have positive values, they can cancel one another out. This leads to the odd conclusion that if a particle can reach a point by two paths rather than one, it might never go there. The nice thing about path integrals, however, is that the method carries over to the standard model of particle physics, but for that we must also include all interactions that the particles could make along the way, such as creating pairs of other particles that then disappear again.

With the path-integral, you can keep pushing on to shorter distance scales, and eventually everything comes down to the twenty-five elementary particles and four forces: electromagnetism, the strong and weak nuclear forces, and gravity. The first three have quantum properties. Physicists haven't yet succeeded in turning gravity into a quantum theory too.

o o o

If I got to nominate the most beautiful, most powerful, and most unifying principle, it would be the principle of least action. But, ack, we still have these twenty-six constants! Can't we find a simpler description for the universe? Maybe one that'll do with only six constants? Or no constants at all?

Physicists have certainly tried. They have put forward many approaches to unified theories in which they calculate some of these constants from other assumptions, or at least predict two of them from one common principle. There have been quite a few attempts, for example, to predict the amount of dark matter together with that of dark energy, or find patterns in the masses of elementary particles. The problem with these ideas is that, so far, they've been more complicated than just writing down the constants. They lack explanatory power.

Indeed, one can interpret multiverse theories as attempts at reducing the number of constants too. If the probability distribution over the different universes would allow us to calculate the observed constants as the most likely values, and if the probability distribution were simpler than just postulating the constants themselves, that would be an improvement over the current theories. If that was possible, however, one could just take the probability distribution as an equation from which to determine the constants. One still wouldn't need the other universes. In any case, no one has so far managed to come up with anything simpler than those twenty-six constants.

A particularly controversial attempt to explain the constants of nature is the strong **anthropic principle**, which says the constants are what they are *because* the universe gave rise to life. Most scientists dismiss this idea out of hand, but I think it's worth thinking about.

First, though, we have to distinguish the strong anthropic principle from the weak anthropic principle. The latter says the constants of nature must allow for the existence of life; otherwise, we wouldn't be here to talk about it. The weak anthropic principle is merely an observational constraint on our theories. It sounds funny because the observation that is used to constrain the theories is self-referential, namely the fact that we are here to make observations at all. But other than that, it's standard scientific argumentation. For example, you can use the observation that you're still reading this book to deduce that there's oxygen in the air around you. That's a weak anthropic constraint, if not exactly a groundbreaking insight.

But weak anthropic constraints can be useful. Fred Hoyle, for example, famously used the fact that life on Earth is carbon-based to deduce that all the carbon must have come from somewhere. This led him to conclude that nuclear fusion in stars must work differently from how physicists at the time thought it did. He was right.

The strong anthropic principle, however, makes a much bigger claim: that the existence of life today is the reason the universe is this way and not any other. Life doesn't just constrain the constants; it explains them. At least that's the idea.

We already know that the strong anthropic principle, taken at face value, is wrong. That's because physicists have found several ways in which the constants of nature could be substantially different and still give rise to chemistry complex enough to create life. Of course, physicists can't calculate structures all the way up to biology, so, strictly speaking, they didn't show that life is possible for other constants of nature. But it's plausible that chemistry as complex as our own can give rise to structures as complex as our own. A recent counterexample to the strong anthropic principle is that the nuclear fusion process, which Hoyle argued must exist because we need all that carbon, isn't necessary for life. There are different values of the fundamental constants that enable other fusion processes that also produce carbon. For the evolution of life, the result of those other processes is pretty much indistinguishable from Hoyle's because it doesn't matter for cells how the carbon they need was produced. I leave you references to some further examples in the endnotes.

But the much bigger problem with the strong anthropic principle is that it's hard to see how it could ever have explanatory power. It's a practical problem: *life* is difficult to define, it is even more difficult to quantify, and therefore you can't calculate anything from the statement "The universe contains life." Those twenty-six constants and their equations are dramatically simpler. Physics ftw!

However, this brings up the question whether there is a different, simpler criterion that our universe fulfills, which is optimal for exactly the values of the constants we observe, and no others. Such a function would quantify in just what way our universe is "the best of all possible worlds," and that would allow us to calculate the constants.

What might such a criterion be? One idea, pioneered by Lee Smolin in his theory of *cosmological natural selection,* is that our universe is really good at producing black holes. According to Smolin, black holes create new universes inside themselves, and new universes randomly receive new constants of nature. If universes can reproduce and give rise to new combinations of constants, then, in the end, the most likely universes are those that produce the most offspring, i.e., that create the most black holes.

The assumptions that (a) black holes give birth to new universes and that (b) the constants of nature can change in that process are both highly speculative and supported neither by our current theories nor by actual evidence. But we don't need these assumptions. We can instead just think of the number of black holes as a function that quantifies how good our universe is. Is our universe with its twenty-six constants "the best" to make black holes?

Let us have a quick look at how this works. Most black holes are formed by stellar collapse, but to form a black hole, a star must be massive enough. Our Sun, for example, cannot form a black hole, because it's too small (it'll most likely become a red giant). This means the number of black holes depends on how efficiently massive stars form from the hydrogen clouds that the early universe's hot plasma left behind.

Just changing the strength of gravity doesn't change the number of black holes; it changes the average mass of stars, but not the fraction of these stars that collapse to black holes. But what about the cosmological constant? If we change it, what happens to the number of black holes?

As we've seen earlier, if the cosmological constant increases, the universe expands faster, and that makes it more difficult for galaxies to form. Most star formation happens in galaxies, so if the cosmological constant were larger, we'd have fewer stars and then fewer black holes. If, on the other hand, the cosmological constant were smaller, the universe would expand more slowly, and galaxies would be more likely to merge. In these mergers, the gas from which stars form is distributed in the now bigger galaxies. This makes star formation less efficient, and again we get fewer stars and then fewer black holes. Our cosmological constant seems to be "the best" for making black holes.

Smolin has put forward similar arguments for several other constants of nature, showing that if you change them away from the values they have, the number of black holes goes down. I have to say, for such a simple idea, it works remarkably well. But this procedure also demonstrates the limits of the approach. We don't know how to write down "number of black holes in the universe" in a simple way, so we can't calculate the constants of nature from it. We can merely see what happens if we change one of those constants at a time. In the end, we're better off just postulating the constants of nature again.

A recurring related idea is to use the growth of complexity as the property that our universe is "the best" at. But, as with "life," the problem is that "complexity" is a vague criterion, and no one currently knows how to quantify it. The best idea I have heard so far is that of David Deutsch, who has conjectured that the laws of nature are so that they will give rise to certain types of computers. It's a good idea, because it can be made formally precise, and I am curious to see what will come out of it.

These ideas have in common the feature that to find better descriptions of nature, they don't follow the trodden path of reductionism: toward shorter distance. Instead, they decouple ontological reductionism from theory reductionism, positing that a better theory might be found on large scales. I think this change of direction has much promise. It's the only approach I know of that might allow us to overcome the problem of initial conditions, which I mentioned in chapter 2.

Will We Ever Know It All?

Physicists certainly have a knack for coming up with imaginative nomenclature: *many-worlds, black holes, dark matter, wormholes, grand unification,* and the *Big Bang. Theory of everything* is another one of those imaginative terms. This conjectured theory of everything would finally explain it all—the elementary particles, the forces among them, and the constants of nature—without leaving further questions. It would be a new and improved fundamental formula, combining both the standard model of particle physics and Einstein's general relativity into one harmonious whole.

However, such a theory of everything, if it exists, would *not* explain everything. That's because, as we discussed in chapter 4, in most areas of science, emergent (effective) theories are better explanations. If we ever find this theory of everything, therefore, we can close the department of particle physics, but we'll continue doing materials science and biomedicine.

Closing the department of particle physics might be worth the effort. But is there such a thing as a theory that leaves no questions unanswered?

Well, one way to get a theory that answers all our questions is to stop asking questions. I'm only partly joking. If we'd stopped doing science two centuries ago and had just called it a day, particle physicists would not be asking now why the mass of the Higgs boson is what it is. I don't want to say that would have been a good decision; I just want to illustrate that whether a theory explains "everything" depends on how much we know, and *want* to know, about nature. If we had a theory that explained everything today, we'd never be sure it'd still explain everything tomorrow.

Even leaving aside the odds that future discoveries may force us to one day revise any purported theory of everything, the idea that a theory answers all questions is itself incompatible with science. Science requires that we formulate different hypotheses for how nature works. We keep the ones that agree with observations and toss the rest. Nevertheless, there are many theories whose "only" problem is that they don't describe what we observe.

Take the theory that the universe is a perfect, empty, two-dimensional sphere. That's not much of a theory, you might say, and I'd agree. But what's its problem? It's not that there is something wrong with the theory in and of itself; there's not much there to be wrong. It's just that it doesn't describe what we observe. It has nothing to do with the universe we actually inhabit.

There are infinitely many consistent theories like this that don't describe what we observe, but one is enough to see the problem: we need the requirement that a theory explains observations to select one theory over others. And this means even the best theory, the one with the highest explanatory power, will answer some questions simply with "because it explains what we observe." Without that, we can't get rid of all the other pretty, simple, and consistent, but empirically inadequate, theories.

A different way to put this is that we can't bootstrap a specific theory for our specific universe out of unspecific math. There is lots of math that just doesn't describe what we see. We select some of that math just because it works.[a] So even if we had a theory of everything, science alone would never explain why that particular theory is the one.

>> THE BRIEF ANSWER

We have no reason to think the universe was made especially for us, or for life in general. It is, however, possible that our current theories are missing something essential about how the laws of nature give rise to complexity in our universe. Maybe the fact that this growth is possible at all will one day give rise to better explanations, flying in the face of reductionism. Still, no scientific theory will ever be able to answer all questions. That's because, for it to be scientific, it must have been selected by its success in explaining observations, but then it will necessarily bounce back some questions with the answer "because it explains what we observe."

---

[a]Tegmark's mathematical universe doesn't change anything about this, because to explain what we observe, you'd still have to specify where we are in the mathematical universe, which is equivalent to having to choose the math that describes our universe.

Chapter 8

DOES THE UNIVERSE THINK?

Size Matters

According to the most recent observations from the Hubble Space Telescope, our universe contains at least 200 billion galaxies. These galaxies are not uniformly distributed—under the pull of gravity, they lump into clusters, and the clusters form superclusters. Between these clusters of different sizes, galaxies align along thin threads, the *galactic filaments*, which can be several hundred million light-years long. Galactic clusters and filaments are surrounded by voids that

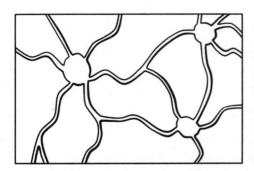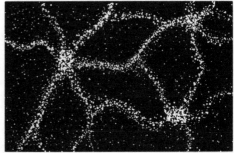

Figure 13: Sketch of neurons (left) and cosmic filaments (right).

contain very little matter. Altogether, the cosmic web looks somewhat like a human brain (figure 13).

To be more precise, the distribution of matter in the universe looks a little like the *connectome*, the network of nerve connections in the human brain. Neurons in the human brain, too, form clusters, and they connect by axons, long nerve fibers that send electrochemical impulses from one neuron to another.

The resemblance between the human brain and the universe is not entirely superficial; it was rigorously analyzed in a 2020 study by the Italian researchers Franco Vazza (an astrophysicist) and Alberto Feletti (a neuroscientist). They calculated how many structures of different sizes are in the connectome and in the cosmic web, and reported "a remarkable similarity." Brain samples on scales below about 1 millimeter and the distribution of matter in the universe up to about 300 million light-years, they find, are structurally similar. They also point out that, "strikingly," about three fourths of the brain is water, which is comparable to the three fourths of the universe's matter-energy budget that's dark energy. In both cases, the authors note, these three fourths are mostly inert.

Could it be, then, that the universe is a giant brain in which our galaxy is merely one neuron? Maybe we are witnessing its self-reflection while we pursue our own thoughts. Unfortunately, this idea flies in the face of physics. Even so, it's worth looking at, because understanding why the universe can't think teaches an interesting lesson about the laws of nature. It also tells us what it would take for the universe to think.

In a nutshell, the universe can't think, because it's too big. Remember that Einstein taught us there is no absolute rest, so we can speak of the velocity of one object only relative to another. This is not the case for sizes. It's not only relative sizes that matter. It's absolute sizes that determine what an object can do.

Take, for example, an atom and a solar system. At first sight, the two have a lot in common. In an atom, the negatively charged electrons are attracted to the positively charged nucleus by the electric force. The strength of this force approximately falls with the familiar $1/R^2$ law, where R is the distance between the electron and the nucleus. In a solar system, a planet is attracted to its sun by the gravitational force. Strictly speaking, that'd be described by general relativity, but a sun's gravity can be well approximated by Newton's $1/R^2$ law, where R is the distance between the planet and the sun. Atoms and solar systems are really quite similar in this regard. Indeed, this is how many physicists thought about atoms at the beginning of the twentieth century; it's basically how the 1913 Rutherford-Bohr model works.

But we know today that atoms aren't little solar systems (figure 14). Electrons aren't small balls that orbit around the nucleus; they have pronounced quantum properties and must be described by wave functions. An electron's position is highly uncertain within the atom, and its probability distribution is a diffuse cloud that takes on symmetrical shapes, called *orbitals*. The electrons' energy in the orbitals comes in discrete steps—it is *quantized*. This quantization gives rise to the regularities we find in the periodic table.

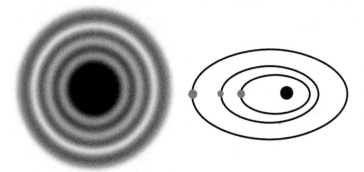

Figure 14: *Atomic energy levels don't work like solar systems. Left: The probability to find an electron in the third shell around the atomic nucleus. Shells are three-dimensional; in the simplest case, they are spheres. The brighter the shade, the higher the probability. The atomic nucleus is in the center but not depicted. Right: Sketch of planetary orbits. Orbits are planar and planets are localized on them.*

This doesn't happen for solar systems. Planets in solar systems can be at any distance from the sun. They aren't fluffy probability distributions, the orbits aren't quantized, and there's no periodic table of solar systems. Where does this difference come from?

The major reason solar systems are different from atoms is that atoms are smaller and their constituents lighter. Because of quantum mechanics, everything—a small particle but also a large object—has an intrinsic uncertainty, an irreducible blurriness in position. The typical quantum blurriness of an electron (its *Compton wavelength*) is about 2×10^{-12} meters. It's similar to the size of a hydrogen atom, which is about 5×10^{-11} meters. That these sizes are comparable is why quantum effects play a big role in atoms. But if you look at the typical quantum uncertainty of planet Earth, that's about 10^{-66} meters and entirely negligible compared with our planet's distance from the Sun, which is about 150 million kilometers (93 million miles). The physical properties—size and mass—make a real difference. Scaling up an atom does not give us a solar system. That's just not how nature works. To use the technical term, the laws are not *scale-invariant*.

Why aren't the laws of nature scale-invariant? It's because of those twenty-six constants. They determine which physical processes are important on which scales, and each scale is different.

We see this scale dependence of physics come through in biology. For small animals like insects, friction forces (created by contact interactions) are much more important than for large animals, like us. That's why ants can crawl up walls and birds can fly, whereas we can't. We're just too heavy. A human-size, human-weight ant would be an evolutionary disaster—and still couldn't crawl up walls. It's not the shape that allows small animals to master these feats; it's just that they don't have to fight so much against gravity.

Let us, then, look at how similar the universe actually is to a brain, keeping in mind that those constants of nature make a difference.

The universe expands, and its expansion is speeding up. How fast the expansion speeds up is determined by the cosmological constant, which is the simplest type of dark energy. Brains, in contrast, don't normally expand, unless possibly metaphorically, and they also don't expand along with the universe: the brain is held together by electromagnetic and nuclear forces, which are much stronger than the pull that the cosmological expansion exerts. Even galaxies themselves are held together by their own gravitational pull and don't expand together with the universe. It's only somewhere between the distances of galaxy clusters and filaments that the expansion of the universe wins over and stretches the galactic web.

So if galaxy clusters were the universe's neurons, then they'd be flying apart from one another with ever-increasing relative speed—and they have been doing so for some billion years already. Dark energy may be "inert," as Vazza and Feletti wrote in their paper, but it plays an important role for the structure of the universe. And while the fraction of dark energy in the universe is similar to the fraction of water in the brain, water doesn't expand the brain. (Or if it does, that's bad news.)

The other constant that makes a major difference between the universe and the brain is the speed of light. Neurons in the human brain send about 5 to 50 signals per second. Most of these signals (80 percent) are short-distance, going only about 1 millimeter, but about 20 percent are long-distance, connecting different parts of the brain. We need both to think. The signals in our brain travel at 100 meters per second (225 miles per hour), a million times slower than the speed of light. Before you conclude that's really slow, let me add that pain signals travel even slower, at only about 1 meter per second. I recently bumped my toe on the door while I happened to be looking at my foot. I just about managed to think, "It's going to hurt," before the pain signal actually arrived.

Maybe our universe is smarter than Einstein and has figured out a way to signal faster than light. But let's put aside such speculations for now and stick with established physics. The universe is now some 90 billion light-years in diameter. This means if one side of the hypothetical universe-brain wanted to take note of its other side, that "thought" would take 90 billion years at least to arrive. Sending a single signal to our nearest galaxy cluster/neuron (the M81 Group) would take about 11 million years, even at the speed of light. At most, the universe might have managed about a thousand exchanges between its nearest neurons in its lifetime so far. If we leave the long-range connections entirely aside, that's about as much as our brain does in three minutes. And the capacity of the universe to connect with itself decreases with its expansion, so it'll go downhill from now on.

The bottom line is, if the universe is thinking, it isn't thinking very much. The amount of thinking that the universe could conceivably have done since it came into existence is limited by its enormous size—and size matters. The physics just doesn't work out. If you want to do a lot of thinking, it helps to keep things small and compact.

o o o

There's still the question whether the whole universe might be connected in a way we don't yet understand, a way that allows it to overcome the speed-of-light limit and do some substantial thinking. Such connections are often attributed to entanglement in quantum mechanics, a nonlocal quantum link that can span large distances.

Particles that are entangled share a measurable property, but we don't know which particle has which share until we measure it. Suppose you have a big particle whose energy you know. It decays into two smaller particles: one flies left; one flies right. You know that the total energy must be conserved, but you don't know which of the decay products has which share—they are *entangled*: the information about the total energy is distributed between them. According to quantum mechanics, which of the small particles has which share of the total energy will be determined only when you make a measurement. But once you measure the energy share of one of the small particles, that of the other small particle—which meanwhile could be far away—is also determined, immediately.

That indeed sounds like something you might use to signal faster than light. However, no information can be sent with this measurement, because the outcome is random. The experimenter who measures one of the particles cannot ensure he will get a particular outcome, so he has no mechanism by which to impress information onto the other particle.

The idea that entanglement is an instantaneous connection over long distances is fertile ground for science myths. Two years ago, I took part in a panel discussion together with another author, who'd recently published a book about dinosaurs.[a] Just because, I suppose, the paleontology section is next to physics. The host, in his best attempt to transition from dinosaurs to quantum mechanics, asked me whether the dinosaurs might have been entangled throughout the universe with the meteoroid that spelled their doom.

That transition deserves a prize, but if you look at the physics, the idea makes no sense. First, as we discussed earlier, quantum effects get washed out incredibly quickly for big objects like you and me, dinosaurs, and meteoroids. Really, you can debunk 99 percent of quantum pseudoscience just by keeping in mind that quantum effects are incredibly fragile. You can't cure diseases with quantum entanglement any more than you can build houses from air, and you can't use entanglement to explain the demise of the dinosaurs either.

Maybe more important, entanglement in quantum mechanics is often portrayed as much more mysterious than it really is. While

---

[a] It wasn't Lisa Randall.

entanglement is indeed nonlocal, it is still created locally. If I break apart a cookie and give you one half, then these two halves are nonlocally correlated because their lines of breakage fit together even though they're spatially separated. Entanglement is a nonlocal correlation like that, but it's quantitatively stronger than the cookie correlation.

I don't want to downplay the relevance of entanglement. That quantum correlations are different from their nonquantum counterparts is why quantum computers can do some calculations faster than conventional computers. But the reason for this computational advantage is not that the quantum correlations are nonlocal; it's that entangled particles can do several things at the same time (with the warning that that's a verbal description of mathematics that has no good verbal description).

I believe the major reason so many people think entanglement is what makes quantum mechanics "strange" is that it's almost always introduced together with the Einstein quote "spooky action at a distance." Einstein indeed used this phrase (or its German translation, "spukhafte Fernwirkung") to refer to quantum mechanics. But he didn't use it to refer to entanglement. He was referring instead to the reduction of the wave function. And that is indeed nonlocal—if you think it is a physical process.

Now, most physicists today don't think the reduction of the wave function is physical, but we don't know for sure what is going on. As Penrose pointed out, it's a gap in our understanding of nature. And this is only one of the reasons physicists in the past decades have toyed with the idea of bona fide nonlocality, not just nonlocal entanglement, but actual nonlocal connections in space-time through which information can be sent across large distances instantaneously, faster than the speed of light.

This isn't necessarily in conflict with Einstein's theories. Einstein's special and general relativity don't forbid faster-than-light motion per se. Rather, they forbid accelerating something from below to above the speed of light, because that would take infinite energy. The speed of light is thus a barrier, not a limit.

Neither does faster-than-light motion or signaling necessarily lead to causality paradoxes—the type where someone travels back in time, kills their own grandfather, and is never born and so they can't travel back in time. Such causality paradoxes can occur in special relativity when faster-than-light travel is possible, because an object that moves faster than light for one observer can look as if it's going back in time for another observer. Thus, in special relativity, you always get both together: faster-than-light motion and backward-in-time motion, and that opens the door to causality paradoxes.

In general relativity, however, causality problems can't occur, because the universe expands and that fixes one direction of time as forward. This forward-in-time direction is related to the forward-in-time direction from entropy increase. The exact relation between them is still somewhat unclear, but that's not so relevant here. What's relevant is only that the universe arguably has a forward direction in time. For this reason, nonlocality and faster-than-light signaling are neither in conflict with Einstein's principles nor necessarily unphysical.

Instead, if they existed, that might solve some problems in the current theories—for example, the issue that information seems to get lost in black holes, which creates an inconsistency with quantum mechanics (see chapter 2). A black hole horizon traps light and everything slower than light, but nonlocal connections can cross the horizon. With them, information can escape and the problem is solved. Some physicists have also suggested that dark matter is really a misattribution. There may be only normal matter whose gravitational attraction is multiplied and spread out because of nonlocal connections in space-time.

These are speculative ideas without empirical support, and I can't say I am enthusiastic about them. I mention them just to demonstrate that nonlocal connections spanning the universe have been seriously considered by physicists. They're far-fetched, all right, but not obviously wrong.

Where might such nonlocal connections come from? One possibility is that they were left behind from geometrogenesis. As we briefly discussed in chapter 2, geometrogenesis is the idea that the universe is fundamentally a network that merely approximates the smooth space of Einstein's theories. However, when the geometry of space-time was created from the network in the early universe, defects might have been left in it. This means, as Fotini Markopoulou and Lee Smolin pointed out in 2007, space would today be sprinkled with nonlocal connections (figure 15).

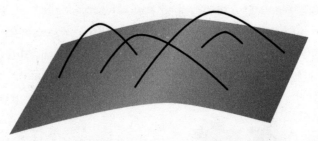

Figure 15: Nonlocal connections (black) in space (gray) work like miniature wormholes. No time passes as one travels from one end to another.

You can think of those nonlocal connections as tiny wormholes, shortcuts that connect two normally distant places. These nonlocal connections would be too small for us, or even elementary particles, to go through. They'd have a diameter of merely 10^{-35} meters. But they would tightly connect the geometry of the universe with itself. And there'd be loads of these connections. Markopoulou and Smolin estimate our universe would contain about 10^{360} of them. The human brain, for comparison, has a measly 10^{15}. And because these connections are nonlocal anyway, it doesn't matter that they expand with space.

I have no particular reason to think these nonlocal connections actually exist, or that, if they existed, they'd indeed allow the universe to think. But I can't rule this possibility out either. Crazy as it sounds, the idea that the universe is intelligent is compatible with all we know so far.

Is There a Universe in Each Particle?

In the previous section, we saw that the laws of nature are not scale-free; that is, physical processes change with the size of objects. But there is a weaker form of scale-free-ness that you may be familiar with: fractals. Take, for example, the Koch snowflake. It's generated by adding smaller equilateral triangles to equilateral triangles, as shown in figure 16a. The shape you get if you continue adding triangles indefinitely is a fractal; the area is finite, but the length of the perimeter is infinite.

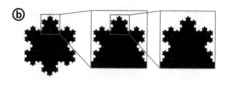

Figure 16a: The Koch Snowflake is created by adding smaller equilateral triangles on triangles, infinitely often.

Figure 16b: The triangular patterns on the Koch Snowflake repeat exactly for the right level of zoom.

The Koch snowflake is not scale-free; it changes if you zoom in on one of its corners. But at the right levels of zoom, the pattern will repeat exactly (figure 16b). If you keep on zooming, it will repeat again and again. We say the Koch snowflake has *discrete scale-invariance*. The pattern repeats only for certain values of the zoom, not all of them. If our universe is not scale-free, then can it have discrete scale-invariance instead, so each particle harbors a whole universe? Maybe there are literally universes inside us. The mathematician and entrepreneur Stephen Wolfram has speculated about this: "[Maybe] down at the Planck scale[a] we'd find a whole civilization that's setting things up so our universe works the way it does."

For this to work, structures wouldn't need to repeat exactly during the zoom. The smaller universes could be made of different elementary particles or have somewhat different constants of nature. Even so, the idea is difficult to make compatible with what we already know about particle physics and quantum mechanics.

To begin with, if the known elementary particles contain mini-universes that can come in many different configurations, then why do we observe only twenty-five different elementary particles? Why aren't there billions of them? Worse, simply conjecturing that the known particles are made up of smaller particles—or are made up of galaxies containing stars containing particles, etc.—doesn't work. The reason is that the masses of the constituent particles (or galaxies or whatever) must be smaller than the mass of the composite particle because masses are positive and they add up. This means the new particles must have small masses.

But the smaller the mass of a particle, the easier it is to produce in particle accelerators. That's because to produce the particle, the energy in the particle collision has to reach the energy-equivalent of the particle's mass ($E = mc^2$!). Particles of small mass are thus usually the first to be discovered. Indeed, if you look at the order in which

---

[a]That's distances of about 10^{-35} meters.

elementary particles were discovered historically, you'll see that the heavier ones came later. This means if each elementary particle were made up of smaller things, we'd long ago have seen them.

One way to get around this problem is to make the new particles strongly bound to one another, so it takes a lot of energy to break the bonds even though the particles themselves have small masses. This is how it works for the strong nuclear force, which holds quarks together inside protons. The quarks have small masses but are still difficult to see, because you need a lot of energy to tear them apart from one another.

We don't have evidence that any of the known elementary particles are made up of such strongly bound smaller particles. Physicists have certainly thought about it, though. Such strongly bound particles that could make up quarks are called *preons*. But the models that have been proposed for this[a] run into conflict with data obtained by the Large Hadron Collider, and by now most physicists have given up on the idea. Some sophisticated models are still viable, but in any case, with such strongly bound particles, you cannot create something that resembles our universe. To get structures similar to what we observe, you'd need an interplay of both long-distance forces (like gravity) and short-distance forces (like the strong nuclear force).

Another way the mini-universes might be compatible with observation would be if the particles they're made up of interacted only very weakly with the particles we know already: they'd just pass through normal matter. In that case, producing them in particle colliders could also be unlikely, and they might hence have escaped detection. This is why the elementary particles called *neutrinos*, even though they have small masses, were discovered later than some of the heavier particles. Neutrinos interact so rarely that most of them go through detectors

---

[a]They're called *technicolor models*.

instead of leaving a signal. However, if you want to make a mini-universe from such weakly interacting, low-mass particles, this creates another problem. They should have been produced in large amounts in the early phase of our universe (as, indeed, neutrinos were), and we should have found evidence for that. Alas, we haven't.

As you see, it isn't easy to come up with ways to build the known elementary particles from something else—other particles or microscopic galaxies—without running into conflict with observations. This is why the standard model of particle physics has kept up for so long.

There is another problem with the idea of putting new particles inside the already known ones, and that is Heisenberg's uncertainty relation. In quantum mechanics, the less mass a particle has, the more difficult it is to keep the particle confined in small regions of space, like inside another elementary particle. If you try to make a mini-universe by stuffing a lot of new, low-mass particles into a known elementary particle, they'll just escape by quantum tunneling.

You can circumvent this problem by conjecturing that the inside of our elementary particles has a large volume. Like the TARDIS in *Doctor Who*, they might be bigger on the inside than they look from the outside. Sounds crazy, I know, but it's indeed possible. That's because in general relativity we can curve space-time so strongly that it'll form bags (figure 17). These bags can have a small surface area—i.e., look small from the outside—but have a large volume inside. The physicist John Wheeler (who introduced the terms *black hole* and *wormhole*) called them "bags of gold." (It was one of his less catchy phrases.)

The problem is, they are unstable—the opening will close off, giving rise to either a black hole or a disconnected *baby universe*. We'll

Figure 17: Wheeler's "bags of gold," aka baby universes, look small from the outside but are big on the inside.

talk about those baby universes in the next interview, but because they don't stay in our space, they can't be elementary particles. And if elementary particles were black holes, they'd evaporate and also disappear pretty much immediately. Not only is this something we've never seen elementary particles doing, but it's also a process that'd violate conservation laws we know to be valid. Or, if you'd managed to find a way to prevent evaporation, they could merge to larger black holes, which is incompatible with the observed behavior of elementary particles.

Maybe there's a way to overcome all these problems, but I don't know one. I therefore conclude that the idea that there are universes inside particles is incompatible with what we currently know about the laws of nature.

Are Electrons Conscious?

It's time to talk about *panpsychism*. That's the idea that all matter—animate or inanimate—is conscious; we just happen to be somewhat more conscious than carrots. According to panpsychism, consciousness is everywhere, even in the smallest elementary particles. This idea has been promoted, for example, by the alternative medicine advocate Deepak Chopra, the philosopher Philip Goff, and the neuroscientist Christof Koch. As you can tell already from this list of names, it's a mixed bag. I'll do my best to sort it out.

First let us note that in the entire history of the universe, not a single thought has been thought without having come about through physical processes; hence, we have no reason to think consciousness (or anything else, for that matter) is nonphysical. We don't yet know exactly how to define *consciousness*, or exactly which brain functions are necessary for it, but it's a property we observe exclusively in physical systems. Because, well, we observe only physical systems. If you think your own thoughts are an exception to this, try thinking without your brain. Good luck.

Panpsychism has been touted as a solution to the problem of dualism, which treats mind and matter as two entirely separate things. As I mentioned earlier, dualism isn't wrong, but if mind is separate from matter, it has no effect on the reality we perceive; hence, it's clearly an ascientific idea. Panpsychism aims to overcome this problem by declaring consciousness fundamental, a property that is carried by any kind of matter—it's everywhere.

In panpsychism, every particle carries *proto-consciousness* and has rudimentary experiences. Under some circumstances, like in your brain, the proto-consciousnesses combine to give proper consciousness. You will see immediately why physicists have a problem with this idea. The fundamental properties of matter are our terrain. If there were a way to add or change anything about them, we'd know.

I realize that physicists have a reputation of being narrow-minded. But the reason we have this reputation is that we tried the crazy stuff long ago, and if we don't use it today, it's because we've understood that it doesn't work. Some call it narrow-mindedness; we call it science. We have moved on. Can elementary particles think? No, they can't. It's in conflict with evidence. Here's why.

The particles in the standard model are classified by their properties, which are collectively called *quantum numbers*. The electron, for example, has an electric charge of −1, and it can have a spin value of +½ or −½. There are a few other quantum numbers with complicated names, such as the *weak hypercharge*, but exactly what they're called is not so important. What's important is that there are a handful of those quantum numbers and they uniquely identify the types of elementary particles.

If you then calculate how many new particles of certain types are produced in a particle collision,[a] the result depends on how many variants of the produced particle exist. In particular, it depends on the different values the quantum numbers can take. That's because this is quantum mechanics, and so anything that *can* happen *will* happen. If a particle exists in many variants, you'll therefore produce them all, regardless of whether or not you can distinguish them.

Now, if you want electrons to have any kinds of experiences—however rudimentary they might be—then they must have multiple different internal states. But if that were so, we'd long ago have seen it, because it would change how many of these particles are created in collisions. We didn't see it; hence, electrons don't think, and neither do any other elementary particles. It's incompatible with data.

There are some creative ways you can try to wiggle out of this conclusion, and I've suffered through them all. Some panpsychists try to argue that to have experiences, you don't need different internal states; proto-consciousness is just featureless stuff. But then claiming that particles have "experiences" is meaningless. I might as well claim that eggs have karma, just that you can't see karma and it has no properties either.

Next you can try to argue that maybe we don't see the different internal states in elementary particles, and they become relevant only in large collections of particles. That doesn't solve the problem, though, because now you'll have to explain how this combination happens. How do you combine featureless proto-consciousness to something that suddenly has features? Philosophers call it the *combination problem* of panpsychism, and, yeah, it's a problem. In fact, if proto-consciousness is physically featureless, it's exactly the same problem as trying to understand how elementary particles combine to create conscious systems.

---

[a]The *multiplicity*.

Finally (and this is what my discussions on the topic usually come down to), you can just postulate that proto-consciousness doesn't have *any* measurable properties, and its only observable consequence is that it can combine to what we normally call consciousness. And that's fine in the sense that it doesn't conflict with evidence. But now you just have a weird version of dualism in which unobservable conscious-stuff is splattered all over the place. It's by construction both useless and unnecessary to explain what we observe; hence, it's ascientific again.

In brief, if you want consciousness to be physical "stuff," then you'll have to explain how its physics works. You can't have your cake and eat it too.

o o o

Now that I've told you why panpsychism is wrong, let me explain why it's right.

The most reasonable explanation of consciousness, it seems to me, is that it's related to the way some systems—like brains—process information. We don't know exactly how to define this process, but this almost certainly means that consciousness isn't binary. It's not an on-off, either-or property, but gradual. Some systems are more conscious, others less, because some process more information, others less.

We don't normally think about consciousness that way because, for everyday use, a binary classification is good enough. It's like how, for most purposes, separating materials into conductors and insulators is good enough, though, strictly speaking, no material is perfectly insulating.

There has to be a minimum size for systems to be conscious, however, because you need to have something to process information with. An object that is indivisible and internally featureless—like an electron—can't do that. Just exactly where the cutoff is, I don't know. I don't think anyone knows. But there has to be one, because the properties of elementary particles have been measured very accurately already and they don't think—as we've just discussed.

This notion of panpsychism is different from the previously discussed one because it does not require altering the foundations of physics. Instead, consciousness is weakly emergent from the known constituents of matter; the challenge is to identify under exactly which circumstances. That's the real "combination problem."

There are various approaches to such physics-compatible panpsychism, though not all advocates are equally excited about adopting the name. The aforementioned Christof Koch is among those who have embraced the label *panpsychist*. Koch is one of the researchers who support *integrated information theory*, IIT for short, which is currently the most popular mathematical approach to consciousness. It was put forward by the neurologist Giulio Tononi in 2004.

In IIT, each system is assigned a number, Φ (Greek capital phi), which is the *integrated information* and supposedly a measure of consciousness. The better a system is at distributing information while it's processing the information, the larger the phi. A system that's fragmented and has many parts that calculate in isolation may process lots of information, but this information is not integrated, so phi is small.

For example, a digital camera has millions of light receptors. It processes large amounts of information. But the parts of the system don't work much together, so phi is small. The human brain, on the other hand, is very well connected and neural impulses constantly travel from one part to another, so phi is large. At least that's the idea. But IIT has its problems.

One problem with IIT is that computing phi is ridiculously time-consuming. The calculation requires that you divide up the system you're evaluating in every possible way and then calculate the connections between the parts. This takes an enormous amount of computing power. Estimates show that even for the brain of a worm with only three hundred synapses, calculating phi with state-of-the-art computers would take several billion years. This is why measurements of phi that have actually been done in the human brain used incredibly simplified definitions of integrated information—for example, by calculating connections merely between a few big parts, not between all possible parts.

Do these simplified definitions at least correlate with consciousness? Well, some studies have claimed they do. Then again, others have claimed they don't. The magazine *New Scientist* interviewed Daniel Bor from the University of Cambridge and reported, "Phi should decrease when you go to sleep or are sedated via a general anesthetic, for instance, but work in Bor's lab has shown that it doesn't. 'It either goes up or stays the same,' he says."

Yet another problem for IIT, which the computer scientist Scott Aaronson has called attention to, is that one can think of rather trivial systems that solve some mathematical problem but that distribute information during the calculation in such a way that phi becomes very large. This demonstrates that phi in general says nothing about consciousness.

There are some other measures for consciousness that have been proposed: for example, the amount of correlation between activity in different parts of the brain, or the ability of the brain to generate models of itself and of the external world. Personally, I am highly skeptical that any measure consisting of a single number will ever adequately represent something as complex as human consciousness, but this isn't so relevant here. What's relevant is that we can scientifically evaluate how well measures of consciousness work.

I have to add some words about Mary's room, because people still bring it up to me in the attempt to prove that perception isn't a physical phenomenon. Mary's room is a thought experiment put forward by the philosopher Frank Jackson in 1982. He imagines that Mary is a scientist who grows up in a black-and-white room, where she studies the perception of color. She knows everything there is to know about the physical phenomenon of color and the brain's reaction to color. Jackson asks, "What will happen when Mary is released from her black-and-white room or is given a color television monitor? Will she learn anything or not?"

He goes on to argue that Mary learns something new upon perceiving color herself, and that therefore the sensation of color is not the same as the brain state of the perception. Instead, the mind has a nonphysical aspect—the *qualia*.

The flaw in this argument is that it confuses knowledge about the perception of color with the actual perception of color. Just because you understand what the brain does in response to certain stimuli (color perception or other), it doesn't mean your brain has that response. Jackson himself later abandoned his own argument.

Fact is, scientists can today measure what goes on in the human brain when people are conscious or unconscious, can create experiences by directly stimulating the brain, can literally read thoughts, and have taken first steps to develop brain-to-brain interfaces. There is so far zero evidence that anything about human perception is nonphysical.

I don't find that surprising. The idea that consciousness can't be scientifically studied because it's a subjective experience never made sense, because one's own subjective experience is all any scientist has ever had to work with. They might have believed it's objective, all right, but in the end, it was all inside their head. And that will remain so unless, maybe, we one day solve the solipsism problem by actually connecting brains.

The advice of philosophers of science is certainly still needed in consciousness research to sort out what properties a satisfactory definition of *consciousness* must fulfill, what questions it can answer, and what counts as an answer. But the study of consciousness has left the realm of philosophy. It is now science.

>> *THE BRIEF ANSWER*

Going by the currently established laws of nature, the universe can't think. However, physicists are considering that the universe has many nonlocal connections because that could solve several problems in the existing theories. It's a speculative hypothesis, but if it's correct, the universe might have enough rapid-communication channels to be conscious. However, the idea that there are universes inside particles and that particles are conscious are both either in conflict with evidence or ascientific. Because consciousness quite possibly isn't a binary variable, some versions of panpsychism are compatible with physics.

CAN WE CREATE A UNIVERSE?
An Interview with Zeeya Merali

If you are a regular reader of popular-science articles about physics, you have almost certainly come across Zeeya Merali's writing. She has written for *Scientific American, New Scientist, Discover,* and *Nature,* to name just a few. She also worked with the BBC and NOVA on their science coverage. Zeeya has a knack for covering even the most speculative ideas without falling for cheap sensationalism. She is one of my favorite writers.

Zeeya and I earned our PhDs around the same time—she in 2004, I in 2003—but while I never quite completed the step from research to writing and ended up neither here nor there, Zeeya successfully switched to science journalism after receiving her degree. She also does much of the public outreach for the Foundational Questions Institute, of which I am a member, and so we have run into each other a few times over the years. In 2017, Zeeya published her first book, *A Big Bang in a Little Room: The Quest to Create New Universes,* about physicists' quest to figure out how to create a universe—and maybe one day actually do it.

Remember that according to the currently most popular theory about the origin of our universe—inflation—everything we see around us came out of a quantum fluctuation of the hypothetical inflaton field that permeates the universe. If the field exists, we could produce conditions for a similar creation event in the laboratory, giving birth to a baby universe. This nascent universe would rapidly grow and detach from ours, much as a drop of water pinches off from the tap. From the outside, the newborn universe would briefly resemble a tiny black hole. It would be gone in a fraction of a second. We'd never find out if it had inhabitants or what happened with them.

Creating such a baby universe would require focusing a large amount of energy in a small region of space. This is not possible in the foreseeable future, but it might one day become possible. And keep in mind that physicists don't yet understand the quantum behavior of space and time. If space and time also undergo quantum fluctuations, baby universes could be created spontaneously, without the need to focus large energies. Once again, this is because in a quantum theory everything that can happen will happen, eventually. If space-time *can* give rise to baby universes—and, mathematically, nothing seems to speak against it—then sometime, somewhere, it *will* give rise to one.

o o o

I meant to visit Zeeya in London, but in early 2020 the COVID pandemic put an end to my travel plans. At the time of writing, May 2021, the UK still allows visitors from Germany only with a ten-day quarantine and two regimes of PCR tests. For what would ordinarily have been a day trip, this is not only cumbersome but prohibitively expensive. I hope that by the time you read this, face masks, self-isolation, and closed borders will have begun to fade in collective memory. But right here and right now, with my deadline approaching, I ask Zeeya for an interview on Skype.

After the obligatory check that we can indeed hear each other, I begin with asking her, too, "Are you religious?"

"Well," Zeeya says, "I've just come off a month of fasting for Ramadan, so judge for yourself." And so I proceed to inquire whether she thinks that scientists will really one day create a universe in the laboratory.

"How the heck do I know?" Zeeya says and laughs. "I'm just someone who writes about it. When I went into it, I just thought it was a weird and interesting idea. I loved that it was possible to pose the question, that you could think about it. And it was not just a wild idea—there's a long history about it. Alan Guth has written about it, Andrei Linde has written about it. It came out of them trying to understand something serious about how our universe began. There's a scientific underpinning to it. They and others had shown that making a universe would take a finite amount of energy rather than an infinite amount, and then it becomes an engineering problem, a very sophisticated, futuristic engineering problem, but something that could in principle be done. That was surprising and exciting to me. But practically doable? I doubt it."

According to the most optimistic estimate, creating a new universe would take about 10 kilograms in pure energy ($E = mc^2$ again!). You need this energy to get the new universe to grow. Once it does that, it creates more energy on its own, because an expanding spacetime violates energy conservation.[a]

Now, 10 kilograms doesn't sound like much—until you remember that even the world's largest particle colliders merely collide, well, particles. They work with mass equivalents that are twenty-four orders of

---

[a]Unfortunately, in the present epoch, the effect is so tiny, it isn't of practical use.

magnitude below what'd be required to make a universe, and the temperatures they reach are about ten orders of magnitude too small. If we have the right theory for how our universe began, there is nothing in principle about that event that we couldn't reproduce. But in practice, no one is going to do it anytime soon.

Zeeya tells me, "When I spoke to the people who were really involved in the research on how to create a universe, who'd thought about it for many decades, in their heart they really think one day it will be done—and they very well may be right. Some of them have a very romantic picture of this. But for me, I suppose it was more interesting that it *could* be done."

When she began working on her book, Zeeya says, she approached the topic from the scientific side, asking what it would take to actually make a universe. But her publishers thought that wasn't the most interesting part of the story.

"They were asking me, 'Are you interested in the ethical side, in the religious side, in the moral side?' It was a strange experience," Zeeya recalls. "Because you don't write about this if you write a feature article for a scientific magazine—you write about the science, the intellectual pursuit. But the book publishers said, 'For us this is really the whole essence of the book.' And I thought, 'Hang on a second. They're giving me license to talk about something I'm genuinely interested in but that I'd had trained out of me.' As a scientist and as a science journalist, you don't want to sound flaky. These are taboo subjects that you're not allowed to get into. And I thought, 'Yeah, let me ask those scientists.'"

Zeeya found that scientists were more willing to talk about the not-strictly-scientific aspects of their work than she'd anticipated.

"I'd actually expected the scientists would be embarrassed and not say much," Zeeya says. "So I'd lined up some theologians to talk about the 'flaky' bit. But when I sat down with the scientists, what surprised me was how much they had thought about these questions: If we can make a universe, could our universe also have had a creator? And could you tell? What moral responsibility do you have toward beings that might evolve in your baby universe? And other tangential questions that they had run into inadvertently when researching cosmology and quantum foundations, or even quantum gravity: Does the universe require consciousness? Are we embedded in a larger 'field of consciousness' that encompasses us all? Do we have free will? Things they hadn't really spoken about in public before, and that they sometimes hadn't spoken about to their colleagues. Aspects that were not necessarily religious—sometimes they were atheists, or people I had assumed were atheists, or who identified as agnostic—but that you might categorize as spiritual."

She adds examples. "Andrei Linde, the cosmologist, riffed on whether the cosmos must be observed—whether by some 'superconsciousness' or just an inanimate recording device—in order for time to start ticking in the universe. Ideas that bubbled up when he had been thinking about quantum gravity and the early universe. Alex Vilenkin, one of the physicists who is famed for having shown how the universe could be created from 'nothing' by a quantum fluctuation, clarified that by 'nothing' he meant no space-time and no matter—but he was curious about how the quantum laws originated. So not really from nothing—the physics and math were out there.

"And not only were people willing to talk about it, but they felt a sense of relief that they could talk about these things, because they'd never been allowed to talk about it before. One physicist, Tony Zee, told me how he had been berated by a senior colleague when he'd started asking 'big questions' as a young researcher, and had tended to keep quiet in public about such things since."

Then there was Antoine Suarez. "He works in the foundations of quantum mechanics and on the topic of free will," Zeeya explains.

"He had a very set idea of what quantum mechanics should be like in order to fit with his religious belief. He felt very strongly that quantum mechanics had to be deterministic, that there is no uncertainty there, because God knows everything. And he developed an experiment to prove that that was very much driven by his religious belief."

But the result of the experiment didn't support Suarez's belief that nature is fundamentally deterministic. "He changed his understanding of how God works—what it means for God to know everything—based on the outcome of that experiment," Zeeya says, clearly impressed.

"What happened with the theologians you lined up for the 'flaky bit'?" I inquire.

"I ended up not including them," Zeeya says. "I went to interview them, and they had all the rigor when I asked them about ethics. But in a funny way, it killed the spirit of the book.

"When I had asked the scientists those same questions, I got very heartfelt responses—answers that arose because they are burrowing deep into the core of the science," Zeeya explains. "They said things that were very personal, and expressed confusion and admitted not always knowing what to think, sometimes. I didn't want to then bring in a theologian who would say, 'Actually, this is the right way to think about the ethics of the multiverse, or whatever, and what the scientist says about philosophy and ethics isn't very rational and doesn't make sense, when you think about it logically.' I wanted those scientists to be given a voice because they were in the midst of this, and that gave their words value. I wanted to get across their uncertainty, that this was an ongoing thought process for many of them."

I can see what she means, I think. "I feel like the physics is more timeless than the morals and ethics," I say. "I mean, I don't know what people will make of the ethics and morals in two thousand years, but the math will still be the same."

"Yes," Zeeya says. "And given that, if I wanted to hear anybody's opinions, it would be the voice of the people involved in it. What was interesting, I guess, was that they are people who are incredibly rigorous about the physics, but then are able to get very emotional and philosophical in an unimpeded way. Just like any of the rest of us. They have the same questions and uncertainties. They don't know all the answers, and they're quite open to admitting that. I really wanted to include this in the book, that there is this openness in them, because I think people get the impression that science closes a lot of questions, and these people were going, 'I don't have the answer to that.' And they were humble about that. I wanted to bring this out."

"I often feel like the philosophical and spiritual side is something we don't talk about enough in the foundations of physics," I say. "Even though it's so important for many people in the field."

Zeeya nods. "I don't think people always recognize this in themselves, or maybe they think it's a failing. But I don't think it's a failing. I think it's a very natural part of how we chose to devote our life to certain passions and certain callings."

>> THE BRIEF ANSWER

An expanding universe can make its own energy. This means if we can figure out how our universe began, we might be able to kick-start the growth of a new one. The most popular theory that physicists currently have for the beginning of our universe—inflation—might be incorrect, and even if it is correct, the necessary technology is beyond us for now. But it's possible in principle. I know it sounds crazy, but the idea that we might one day create a universe in the laboratory is consistent with all we know.

Chapter 9

ARE HUMANS PREDICTABLE?

The Limits of Math

Do you remember the scene from *Basic Instinct?* No, not *that* one. I mean the scene where they walk up the stairs and he says, "I'm very unpredictable," but she says "unpredictable" along with him. We're not remotely as unpredictable as we like to think.

Indeed, many aspects of human behavior are fairly easy to predict. Reflexes, for example, short-circuit conscious control for the sake of speed. If you hear a sudden, loud sound, I can predict you'll twitch and your heart rate will spike. Other aspects of human behavior are predictable for group averages; they stem, among other things, from the constraints of economic reality, social norms, laws, and upbringing. Take the unsurprising fact that traffic is usually worse during rush hour. Indeed, mobility patterns are in general 93 percent predictable, according to an analysis of data collected from mobile phone users. I can also predict that in North America, undressing in public will bring you a lot of attention. And that Brits drink tea, watch cricket, and if you have a foreign accent, will inevitably explain to you that the Queen owns the swans in England.[a]

Stereotypes are amusing exactly because humans are, to some extent, predictable. But is human behavior entirely predictable? It is arguably not currently entirely predictable, but that's the boring answer. Is it possible in principle, given all we know about the laws of nature? If you are a compatibilist who believes your will is free because your decisions can't be predicted, must you fear that you will become predictable one day?

In 1965, the philosopher Michael Scriven argued that the answer is no. Scriven claimed there is an "essential unpredictability in human behavior" using what is now called the *paradox of predictability*. It goes like this: Suppose you are given the task of making a decision. For example, I offer you a marshmallow, and you either take it or not. Now let us imagine I predicted your decision and told you about it. Then you could do the opposite, and my prediction would be false! Hence, human behavior has an unpredictable element. It's important that Scriven's argument works even if human behavior is entirely determined by, say, the initial state of the universe. Predictability, it seems, does not follow from determinism.

This conclusion is correct, but it has nothing to do with human behavior in particular. To see why, suppose I write a computer code whose only task is to output YES or NO to the question whether an input number is even. Then I add a clause saying that when the input further contains the correct answer to the first question, the output is the negation of the first answer. That is, the input "44" would result in YES, but the input "44, YES" would result in NO. By Scriven's argument, there'd be something essentially unpredictable about that computer code too.

---

[a]Then they'll apologize and talk about the weather.

Indeed there is, because the prediction for the code's output depends on the input; it's unpredictable without it. There are lots of systems that have this property; for example, your being offered the marshmallow: Your reaction depends on what I say when I offer it. But that doesn't mean it was fundamentally unpredictable; it just means it wasn't predictable from insufficient data. If you'd put the two of us into a perfectly isolated room, then, in a deterministic world, you could predict what both of us would do, and also whether you'd take the marshmallow.

So Scriven's argument doesn't work. But if you've been paying attention, then you know that human behavior is partly unpredictable just because quantum mechanics is fundamentally random. It is somewhat unclear just what role quantum effects play in the human brain, but you don't need those. You could just use a quantum mechanical device—or maybe pull up that Universe Splitter app on your phone—to decide whether or not to take the marshmallow. And I couldn't predict that decision.

I could still, however, predict the probability of your making a particular decision, and I could test how good my predictions are by running the experiment repeatedly, the same way we test quantum mechanics. So, really, when we are asking whether human behavior is predictable, we should be asking more precisely whether the probabilities of decisions are predictable. Insofar as the current laws of nature are concerned, they are—and to the extent they aren't predictable, they aren't under your control.

However, this conclusion seems to be contradicted by some results from computer science. In computer science, there are certain types of problems that are undecidable, meaning it's been mathematically proven that no possible algorithm can solve the problem. Could not something similar go on in the human brain?

One of the best-known undecidable problems is the *domino*

problem, posed by Hao Wang in 1961. Assume you have a set of square tiles. Draw an X on each of them, so you get four triangles on each tile. Then fill each triangle with a color (figure 18). Can you cover an infinite plane with those tiles so the colors of adjacent tiles all fit together if you're not allowed to rotate the tiles or leave gaps? That's the domino problem. It is easy to see that, for certain sets of tiles, the answer is yes—it is possible. But the question Wang posed is: If I give you an arbitrary set of tiles, can you tell me whether it'll tile the plane?

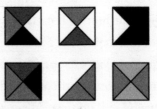

Figure 18: Example for a set of Wang tiles.

This problem, it turns out, is undecidable. One can't write computer code that'll answer the question for all sets of tiles. This was proved in 1966 by Robert Berger, who showed that Wang's domino problem is a variant of Alan Turing's halting problem. The halting problem poses the question whether an input algorithm will finish running at a finite time or continue calculating forever. The problem, Turing showed, is that there is no meta-algorithm that can decide whether any given input algorithm will or won't halt. Likewise, there is no meta-algorithm that can decide whether any given set of tiles will or won't tile the plane.

However, the undecidability of both the domino problem and the halting problem comes from the requirement that the algorithm answer the question for a system of infinite size. In the domino problem, that's all possible sets of tiles; in the halting problem, it's all possible input algorithms. There are infinitely many of both of them.

We already saw this earlier, in chapter 6, when we discussed the question whether some emergent properties of composite systems are uncomputable. These uncomputable properties occur only if some quantity becomes infinitely large, which never happens in reality—certainly not in the human brain.

So if we can't argue that our decisions might be algorithmically undecidable, what about the argument from Gödel's incompleteness theorem that Roger Penrose brought up? Penrose's argument isn't about predictability, but about computability, which is a somewhat weaker statement. A process is computable if it can be produced by a computer algorithm. The current laws of nature are computable, except for that random element from quantum mechanics. If they were uncomputable, though, that'd make space for something new, maybe even unpredictability.

Let us use the approach to Gödel's theorem that Penrose mentioned, which he credits to Stourton Steen. We start with a finite set of axioms and imagine a computer algorithm that generates theorems derived from those axioms, one after another. Then, Gödel showed, there is always a statement, formulated within this system of axioms, that is true but that the algorithm cannot prove to be true. This statement is usually called the *Gödel sentence* of the system.[a] It's constructed so it implicitly states that it's unprovable within the system. Therefore, the Gödel sentence is true exactly because it can't be proven, but its truth can be seen only from outside the system.

It might seem, then, that because *we* can see the truth of the Gödel sentence, whereas the algorithm can't, there's something about human cognition that a computer doesn't have. However, this particular insight about the Gödel sentence is uncomputable only by that particular algorithm. And the reason we can see the truth of this Gödel

---

[a]Though there are infinitely many different statements that can perform this function.

sentence is that we have more information about the system than does the algorithm that's creating all those theorems—we know how the algorithm itself was programmed.

If we gave that information to a new algorithm, then the new algorithm would see the truth of the previous algorithm's Gödel sentence, just as we do. But then we could construct another Gödel sentence for the new algorithm, and another algorithm that recognizes the new Gödel sentence, and so on. Penrose's argument is thus that, because we can see the truth of *any* Gödel sentence, we can do more than any conceivable algorithm.

The problem with this argument is that computer algorithms, suitably programmed, are—for all we can tell—as capable of abstract reasoning as we are. We can't count to infinity any better than a computer, but we can analyze the properties of infinite systems, both countable and uncountable ones. So can algorithms. That way, Gödel's theorem itself has been proven algorithmically. Hence, some algorithms, too, can "see the truth" of all Gödel sentences.

There are a number of other objections that have been raised to Penrose's claim, but most of them likewise come down to pointing out that humans simply wouldn't see the truth of a Gödel sentence without further information—like Gödel's theorem—either. However, I do find quite charming the argument that humans would recognize $\forall x \neg \mathrm{Prf_F}(x, \ulcorner G_F \urcorner)$ as obviously true. That's an idea only a mathematician could come up with.

Could a computer have come up with the proof of Gödel's theorem on its own? That's an open question. But at least for now, Penrose's argument doesn't show that human thought is non-computable.

So far, we haven't found any loophole that would allow human behavior to be unpredictable. But what about chaos? Chaos is deterministic, but just because it's deterministic doesn't mean it can be predicted. Indeed, chaos could be more of a problem for predictability than commonly thought, because of what Tim Palmer dubbed the "real butterfly effect."

The common butterfly effect has it that the time-evolution of a chaotic system is exquisitely sensitive to the initial conditions; the smallest errors (a butterfly flap in China) can make a large difference later (a tornado in Texas). The real butterfly effect, in contrast, means that even arbitrarily precise initial data allow predictions for only a finite amount of time. A system with this behavior would be deterministic and yet unpredictable.

However, while mathematicians have identified some differential equations with this behavior, it is still unclear whether the real butterfly effect ever occurs in nature. Quantum theories are not chaotic to begin with and therefore can't suffer from the real butterfly effect. In general relativity, singularities can prevent us from making predictions beyond a finite amount of time, like inside black holes or at the Big Bang. However, as we discussed earlier, these singularities likely just signal that the theory breaks down and needs to be replaced with something better. And if general relativity is one day completed by a quantum theory, then that, too, cannot have a real butterfly effect.

The best chance for a breakdown of predictability comes—like the "common" butterfly effect—from weather forecasts. In this case, the dynamical law is the Navier-Stokes equation, which describes the behavior of gases and fluids. Whether the Navier-Stokes equation always has predictable solutions is still unknown. Indeed, it is number four on the list of the Clay Mathematics Institute's Millennium Problems.

But the Navier-Stokes equation is not fundamental; it emerges from the behavior of the particles that make up the gas or fluid. And we already know that fundamentally—on the deepest level—the gases are described by quantum theories again, so their behavior is predictable, at least in principle. This does not answer the question whether the Navier-Stokes equation always has predictable solutions, but if it doesn't, it's because the equation does not take into account quantum effects.

So far it seems we have no reason to think human behavior is uncomputable, that human decisions are algorithmically undecidable, or that human behavior might be predictable for only a finite amount of time. Especially in light of the neuron-replacement argument from chapter 4, it full well looks as though we can simulate brains on a computer and therefore predict human behavior.

Physics puts several obstacles in the way, however. Perhaps the most important one is that replacing a neuron is not the same as copying a neuron. If we wanted to predict a human's behavior, we'd first have to produce a faithful model of the person's brain. For that, we'd have to measure its properties somehow and then copy that information into our prediction machine, whatever that might be. However, in quantum mechanics, the state of a system cannot be perfectly copied without destroying the original system. This *no-cloning theorem* makes it provably impossible for people to know exactly what is going on inside your brain, because if they knew, they'd have changed your brain. Therefore, if any relevant details of your thoughts are in quantum format, they are "unknowable" and hence unpredictable.

Quantum effects, however, might not actually matter very much to exactly define the state of your brain. Even if they don't, though, there's another obstacle in the way of predicting human behavior. Our brains are not particularly good at crunching through difficult math problems, but they're remarkably efficient for making complex decisions—while running on only about 20 watts, about the power consumption of a laptop. If you could produce a simulation of a human brain on a computer, it's therefore questionable that it would actually run faster than the brain it's trying to simulate. To use the term coined by Stephen Wolfram, human deliberation might be computable but not *computationally reducible*, and therefore not predictable, in the sense that the calculation may be correct, but too slow.

It's not an implausible conjecture that part of our behavior is computationally irreducible. The human brain was optimized by natural selection over hundreds of thousands of years. If someone wanted to predict it, they'd first have to build a machine capable of doing the same thing, faster. However, for the same reason—that it's been produced by natural selection—it's also unlikely that the human brain is really the fastest way to compute what our brains compute. Natural selection isn't in the business of coming up with the best overall solutions. Solutions just have to be good enough to survive. And if we take into account that a computer would not be required to be as energy efficient as the brain, I suspect it'll be possible to outdo the human brain in speed. But it'll be difficult.

For the same reason, I strongly doubt we will ever derive morals, as Sam Harris has argued, from whatever knowledge we gather about the human brain. Even if it does become possible, it'd just be too time-consuming. It is much easier to just ask people what they think, which is, in a nutshell, what our political, economic, and financial systems do. Or at least what they should do.

In summary: we have no reason to think human behavior is unpredictable in principle, but good reason to think it's very difficult to predict in practice.

AI Fragility

Having discussed the challenges in the way of simulating human behavior, let us talk for a bit about attempts to create artificial general intelligence. In contrast to the artificially intelligent systems we use right now, which specialize in certain tasks—like recognizing speech, classifying images, playing chess, or filtering spam—an artificial general intelligence would be able to understand and learn as well as humans, or even better.

Many prominent people have expressed worries about the aim to develop such a powerful artificial intelligence (AI). Elon Musk thinks it's the "biggest existential threat." Stephen Hawking said it could "be the worst event in the history of our civilization." Apple cofounder Steve Wozniak believes that AIs will "get rid of the slow humans to run companies more efficiently." And Bill Gates, too, put himself in "the camp that is concerned about super intelligence." In 2015, the Future of Life Institute formulated an open letter calling for caution and formulating a list of research priorities. It was signed by more than eight thousand people.

Such worries are not unfounded. Artificial intelligence, like any new technology, brings risks. While we are far from creating machines even remotely as intelligent as humans, it's only smart to think about how to handle them sooner rather than later. However, I think these worries neglect the more immediate problems AI will bring.

Artificially intelligent machines won't get rid of humans anytime soon, because they'll need us for quite some while. The human brain may not be the best thinking apparatus, but it has distinct advantages over all machines we have built so far: It functions for decades. It's robust. It repairs itself. Some million years of evolution optimized not only our brains but also our bodies, and while the result could certainly be further improved (damn those knees), it's still more durable than any silicon-based thinking apparatuses we have created to date. Some AI researchers have even argued that a body of some kind is necessary to reach human-level intelligence, which—if correct—would vastly increase the problem of AI fragility.

Whenever I bring up this issue with AI enthusiasts, they tell me that AIs will learn to repair themselves, and even if they don't, they will just upload themselves to another platform. Indeed, much of the perceived AI threat comes from their presumed ability to replicate themselves quickly and easily, while at the same time being basically immortal. I think that's not how it will go.

It seems more plausible to me that artificial intelligences at first will be few and one of a kind, and that's how it will remain for a long time. It will take large groups of people and many years to build and train artificial general intelligences. Copying them will not be any easier than copying a human brain. They'll be difficult to fix once broken, because, as with the human brain, we won't be able to separate the hardware from the software. The early ones will die quickly for reasons we will not even comprehend.

We see the beginning of this trend already. Your computer isn't like my computer. Even if you have the same model, even if you run the same software, they're not the same. Hackers exploit these differences between computers to track your internet activity. *Canvas fingerprinting*, for example, is a method of asking your computer to render a font and output an image. The exact way your computer performs this task depends on both your hardware and your software; hence, the output can be used to identify a device.

At present, you do not notice these subtle differences between computers all that much (except possibly when you spend hours browsing help forums, murmuring, "Someone must have had this problem before," but turn up nothing). The more complex computers get, the more obvious the differences will become. One day, they will be individuals with irreproducible quirks and bugs—like you and me.

So we have AI fragility plus the trend that increasingly complex hard- and software becomes unique. Now extrapolate this some decades into the future. We will have a few large companies, governments, and maybe some billionaires who will be able to afford their own AI. Those AIs will be delicate and need constant attention from a crew of dedicated humans.

If you think about it this way, a few problems spring up immediately:

1. Who gets to ask questions, and what questions?
This may not be a matter of discussion for privately owned AIs, but what about those produced by scientists or bought by governments? Does everyone get a right to a question per month? Do difficult questions have to be approved by the parliament? Who's in charge?

2. How do you know you are dealing with an AI?
The moment you start relying on AIs, there's a risk that humans will use them to push an agenda by passing off their own opinions as those of the AI. This problem will occur well before AIs are intelligent enough to develop their own goals. Suppose a government uses AI to find the best contractor for a lucrative construction task. Are you sure it's a coincidence that the biggest shareholder of the chosen company is the brother of a high-ranking government official?

3. How can you tell if an AI is any good at giving answers?
If you have only a few AIs, and those are trained for entirely different purposes, it may not be possible to reproduce any of their results. So how do you know you can trust them? It could be a good idea to require that all AIs have a common area of expertise that can be used to compare their performance.

4. How do you prevent the greater inequality, both within nations and between nations, inevitably produced by limited access to AI?

Having an AI to answer difficult questions can be a great advantage, but left to market forces alone, it's likely to make the rich richer and leave the poor even further behind. If this is not something that the "un-rich" want—and I certainly don't—we should think about how to deal with it.

Personally, I have little doubt that an artificial general intelligence is possible. It may become a great benefit for human civilization—or a great problem. It is certainly important to think about what ethics to code in to such intelligent machines. But the most immediate problems we will have with AIs will come from *our* ethics, not theirs.

Predicting Unpredictability

I've spent most of this book discussing what physics teaches us about our own existence. I hope you've enjoyed the tour, but maybe you sometimes couldn't avoid the impression that this is heady stuff that doesn't do much to solve problems in the real world. And so, as we near the end of this book, I want to spend a few pages on the practical consequences that understanding unpredictability may have in the future.

Let us return to the problem of weather forecasting. We are not going to solve the fourth Millenium Problem here, so for the sake of the argument, let us just assume that solutions to the Navier-Stokes equation are indeed sometimes unpredictable beyond a finite time. As I've explained, we already know that the Navier-Stokes equation isn't fundamental; it instead emerges from quantum theories that describe all particles. But, fundamental or not, understanding the properties of the Navier-Stokes equation tells us what we can reasonably hope to achieve by solving it.

If we knew we couldn't improve weather forecasts, because a mathematical theorem said it was impossible, we might, for example, conclude it doesn't make sense to invest huge amounts of money into additional weather measurement stations. Whether the Navier-Stokes equation is fundamentally the right equation doesn't make this investment advice any less sound; it matters only that it's the equation meteorologists use in practice.

This is an oversimplified case, of course. In reality, the feasibility of a prediction depends on the initial state: some weather trends are easy to predict over long periods, others not. But again, understanding what can be predicted in the first place isn't just idle mathematical speculation. It's necessary to know what we can improve, and how.

Let us pursue this thought a little further. Suppose we got really good at doing the weather forecast, so good that we could figure out exactly when the Navier-Stokes equation is about to run into an unpredictable situation. This could then allow us to find out which small interventions in the weather system could change the weather to our liking.

Scientists have indeed considered such weather control, for example, to prevent tropical cyclones from growing into hurricanes. They understand the formation of hurricanes well enough to have come up with methods for interrupting their growth. At present, the major problem is that the weather predictions just aren't good enough to figure out exactly when and where to intervene. But preventing hurricanes, or controlling the weather in other ways, isn't a hopelessly futuristic idea. If computing power continues to increase, we might actually be able to do this within a few decades.

Chaos control also plays a role in many other systems—for example, the plasma in a nuclear fusion plant. This plasma is a soup of atomic nuclei and their disconnected electrons with a temperature of more than 100 million degrees Celsius (180 million degrees Fahrenheit). It sometimes develops instabilities that can greatly damage the containment vessel. If an instability is coming on, therefore, the fusion process must be rapidly interrupted. This is one of the main reasons it is so difficult to run a fusion reactor energy-efficiently.

However, plasma instability is in principle avoidable if we can predict when an unpredictable situation is about to come up, and if we can control the plasma so the situation is averted. In other words, if we understand when a solution to the equations becomes unpredictable, we can use that knowledge to prevent it from happening in the first place.

This is not just the fantasy of a theorist; a recent study looked into exactly this. A group of researchers trained an artificially intelligent system to recognize data patterns that signal an impending plasma instability. They were able to do this with good success, using only data in the public record. A second ahead, they correctly identified an imminent instability in somewhat more than 80 percent of cases; 30 milliseconds ahead, they saw almost all instabilities coming.

Granted, theirs was a hindsight analysis, with no option of active control. However, should we become good enough making such predictions, active control might become possible in the future. An energy-efficient fusion plant, in the end, might be a matter of fine-tuning with advanced machine learning.

A similar consideration applies for a superficially entirely different system that, however, has many parallels to plasma blowups and weather forecasts: the stock market. Today, a whole army of financial analysts makes money by trying to predict the selling and buying of stocks and financial instruments, a task that now includes predicting their competitors' predictions. But every once in a while, even they get caught by surprise. A stock market crashes, vendors panic, everyone blames everyone else, and the world slumps into a recession.

But imagine we could tell in advance when trouble is at the front door; we might be able to close the door.

It's not only unpredictability that we might want to recognize in order to avoid it, but also uncomputability. Take the economic system. It is a self-organized, adaptive system with the task of optimizing the distribution of resources. Some economists have argued that this optimization is partly uncomputable. That's clearly not good, for it means the economic system cannot do its job. Or rather, we as agents in the economic system cannot do our job, because trading does not have the desired result.

Creating an economic system that can actually do the desired optimization (in finite time) has motivated the research line of *computable economics*. And as with unpredictability, what makes impossibility theorems relevant for computable economics is not proving that the solution to a problem (here: how to best distribute resources) is fundamentally uncomputable—it may or may not be—but merely that it is uncomputable with the means we currently have.

In other situations, however, unpredictability is something we may want to trigger rather than avoid, for the same reason that randomness can sometimes be beneficial; for example, to prevent computer algorithms that search for optimal solutions from getting stuck.

Think of the computer algorithm as a device that, when you dump it into a mountainous landscape, will always move uphill. The landscape stands for the possible solutions to a problem and the height for some quantity you want to optimize, say, the accuracy of a prediction. In the end, the computer algorithm will sit on a hill—the *local optimum*—but what you actually wanted to find was the highest hill—the *global optimum* (see figure 19). Adding stochastic noise can prevent this from happening, because the algorithm then has a chance to coincidentally discover a better solution. Counterintuitively, therefore, an element of randomness can improve the performance of mathematical code.

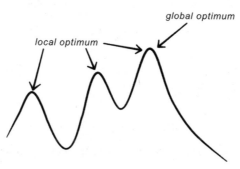

Figure 19: Local vs. global optimum.

In a computer algorithm, randomness can be implemented by a (pseudo) random number generator without drawing on complicated mathematical theorems. But unpredictability might well be useful for optimization in other circumstances. For example, small doses could aid the efficiency of the economic system. Even more interesting, unpredictability might be an essential element of creativity, and thus something that artificial intelligence could draw on in the future.

Already, right now, artificial intelligence is better at discovering patterns in large sets of data than we are. This is about to change science dramatically. Human scientists look for universals—patterns that are robust under changes in the environment and easy to infer. That's how most science has proceeded so far. By using artificial intelligence, we can now look for patterns that are much more difficult to discern. The development of personalized medicine is one consequence, and we will almost certainly see more of this soon. Instead of looking for universal laws, scientists will increasingly be able to track exact dependencies on external parameters—in ecology and biology, for example, but also in social science and psychology. There is vast discovery potential here.

Physicists should take note too. The universal laws they have found might only scratch the surface of so-far-unrecognized complexity. While my colleagues think they are closing in on a final answer, I think we've only just begun to understand the question.

>> THE BRIEF ANSWER

Human behavior is partially predictable, but it's questionable that it'll ever be fully predictable. At the very least, it's going to be extremely difficult and won't happen any time soon. Instead of worrying about simulating human brains, we should pay more attention to who gets to ask questions of artificial brains. Understanding the limits of predictability isn't merely of mathematical interest but is also relevant for real-world applications.

WHAT'S THE PURPOSE OF ANYTHING ANYWAY?

I f you read my previous book, *Lost in Math*, you might have noticed it has a thread in common with this book. It's that I think researchers in the foundations of physics don't reflect enough on what they are doing. In my earlier book, I criticized their use of unscientific methods, as a result of which their research has gotten stuck. In this book, I have pointed out that some of the research they pursue isn't scientific to begin with. Most hypotheses for the early universe, for example, are just complicated stories that are unnecessary to describe anything we observe. The same goes for attempts to find out why the constants of nature are what they are, or theories that introduce unobservable parallel universes. This isn't science. It's religion masquerading as science under the guise of mathematics.

Don't get me wrong. I don't have a problem with people pursuing these ideas per se. If someone finds it valuable for whatever reason, that's fine with me—everyone should be free to exercise their religion. But I want scientists to be mindful of the limits of their discipline. Sometimes the only scientific answer we can give is "We don't know."

It therefore seems likely to me that, in our ongoing process of knowledge discovery, religion and science will continue to coexist for a very long time. That's because science itself is limited, and where science ends, we seek other modes of explanation. As I have laid out in the previous chapters, some of these limits stem from the specific math we currently use (which, for example, requires initial conditions or indeterministic jumps), and they may be overcome as physics advances further. But some limits seem insurmountable to me. Eventually, I think, we will have to accept some facts about our universe without scientific explanation, if only because the scientific method can't justify itself. We may observe that the scientific method works, conclude that it's to our advantage to continue using it, but still never know why it works.

It's not that I want to be nice to religious people for the sake of being nice. To begin with, I'm not exactly known for being nice. But more important, scientists who claim, as Stephen Hawking did, that "there is no possibility of a creator," or as Victor Stenger has, that God is a "falsified hypothesis," demonstrate that they don't understand the limits of their own knowledge. When prominent scientists make such overconfident proclamations, they make me cringe.

Despite all our limitations, however, I have to say we have come a remarkably long way. We are the first species on our planet that has taken evolution into its own hands. No longer are we selected by our natural environment; we shape the environment to our own needs. Whether we are any good at this is another question. Certainly our difficulty in keeping Earth's climate in a comfortably habitable range raises severe doubts as to our cognitive ability to handle complex and partly chaotic systems. Maybe it's because our brains are ill-equipped to understand a system as multifaceted and nonlinear as the climate.

Maybe that means humans will eventually be replaced by a species more capable of using scientific knowledge to control its habitat. Time will tell.

o o o

It isn't only that I think Stephen Jay Gould got it right when he argued that religion and science are two "nonoverlapping magisteria." I will go a step further and claim that scientists can learn something from organized religion. For better or worse, religions have played an important role for big parts of the world population over thousands of years. Religion matters to many people in a way that science doesn't.

Partly this is just because religion has been around longer, but it's also because too many people perceive science as cold, technocratic, and unhumanly rational. It has the reputation of being a killjoy that constrains our hopes and dreams. Of course, it's true that science says flapping your arms won't make you fly. But science has another side: it opens our eyes to possibilities we couldn't previously imagine, much less comprehend. Far from taking away wonder, science gives us more to marvel at. It expands our minds.

The best comparison I can think of is this. I sometimes have lucid dreams—that is, dreams in which I know I am dreaming. I have friends who have tried to trigger lucid dreams but largely failed. I, in contrast, would rather not have those dreams, but it's not as though I can put them up for sale. The main reason I don't like them is that I usually wake up afterward, and that ruins my night rest. But also, they are creepy.

Unlike normal dreams, in which you just accept what you see for what it is, in lucid dreams, I can tell very well that what I'm experiencing isn't real. If I "see" a face, I don't actually see the face. It's more an idea of a face, but when I try to look at it, it isn't there. It's deep in the uncanny valley, but the valley's inside my head. Buildings, objects, and the sky suffer from the same problem. I know they're there, and sometimes I can move them around or change their color, though that doesn't always work. But they lack details. They're ideas of the real thing rather than the real thing. That makes me feel like I am trapped in an old video game, one of those where the walls were perfectly even, infinitely thin planes, but they sometimes didn't fit together at the corners and you'd get stuck between them. Remember that? And while I can fly in my dreams if I want to, there isn't much to see below. It's rather dull, honestly.

I suspect what's happening is that my brain just doesn't store enough details to project the required imagery and experience convincingly. That doesn't surprise me, because how am I supposed to know what flying feels like or a pink sky looks like? And I suck at remembering faces even on the best of days.

The lesson I take away from this is that the world out there is literally richer than we can possibly imagine. We need reality to feed our brain. And this isn't true only for sensorial experience, I think; it's true also for ideas. We get them from our interaction with nature, from our study of the universe—we get them from science. Just as my lucid dreams are pale memories of awake moments, without science our ideas remain pale memories of what we know already.

I wouldn't go so far as Stewart Brand, who claimed that "science is the only news," for science certainly isn't the sole creative discipline that draws inspiration from nature. But science has a way of entirely changing our conception of reality with unanticipated twists. That's why, to me, science is first an inspiration, not a profession. It's a way to make sense of the world and discover genuine novelty. That's a side of science I wish would be celebrated more often.

Scientists can learn from religion that not every get-together needs to come with a teachable lesson. Sometimes we just enjoy the company of like-minded people, want to share experiences, or look forward to a traditional ceremony. Science is severely lacking in such social integration. It's something we can and should improve on. Alongside public lectures, we should offer opportunities for lecture attendees to get to know one another. Instead of panel discussions among prominent scientists, we should talk more about how scientific understanding made a difference for non-experts. Instead of letting researchers answer audience questions, we should listen and learn from those who have been helped through difficult times by scientific insights. A clear view of the night sky, a book on embryology, an online course in psychology, or a lecture on neurophysiology can change lives. I know this because people share such stories with me after lectures, by letter, or on social media. They should be more widely known.

o   o   o

Scientists are often—all too often—required to justify their research by demonstrating practical applications. But we have another reason entirely for our research: the desire to make sense of our own existence. We all have our own approach to sense making, and I have illustrated mine through the examples in this book.

Yet you may ask, "What's the point?" If the universe is just machinery, a set of differential equations acting on initial conditions, and we are but blips of complexity in an uncaring universe, temporarily self-aware conglomerates of particles that will soon be washed away by entropy increase, then why spend time figuring out just exactly how insignificant our existence is? What's the meaning of life if there's no purpose to it?

I don't intend to answer this question for you, not because I don't think there's an answer, but because I believe we all have to find our own answer. Let me just tell you how I personally think about it.

I remember asking my mother, "What's the meaning of life?" when I was maybe fourteen years old. She seemed more tired than surprised by teenage me and, after some consideration, answered that to her the meaning of life is to pass on knowledge to the next generation. My mom, you should know, is a (now retired) high school teacher. Hers was a coherent answer, I thought at the time, but rather lame. Of course a teacher would say that passing on knowledge is the most important thing ever!

Thirty years later, I have come to pretty much the same conclusion. Yes, people also tell me I look like my mother. But while I intended to become a teacher, in the end I didn't, for the simple reason that I don't like to repeat myself. Yet today I would give an answer very similar to my mother's.

You see, for the past two decades, I have been enormously lucky and privileged. Thanks to financial support from governmental funding bodies, private institutions, and individual donors, I have been able to study the fundamental laws of nature and report back to you the conclusions I've arrived at. The feedback I get to my writing, my lectures, and my video channel demonstrates vividly that lots of people care about answers to the same questions I am driven by. They want to know how the universe works.

From a purely economic perspective, my research became possible only because sufficiently many others thought the potential insights would be worth the investment. And yet that's somewhat perplexing, isn't it? There's no financial benefit or selective advantage to knowing what I laid out in this book. One could maybe try to argue that understanding nature is, broadly speaking, good for survival, that nerds are sexy, or that humans spend money on many fads that make no sense at all. But I don't think that cuts it. Basic research isn't just a fad; it's an institutionalized endeavor of advanced societies. We don't study the universe just because we hope to one day travel to other galaxies. Even if we hoped to, and even if we worked toward it, that still wouldn't explain why we care whether time is real or want to know why the constants of nature are what they are.

To me, my personal story is evidence that not only I but many of us have the desire to understand the universe—for no other reason than understanding the universe. Our thirst for knowledge is ubiquitous, in both individuals and societies. We want to understand, partly because understanding is useful, but also, I think, out of a primary need to make sense of ourselves and our place in this world.

Maybe, then, the universe is evolving toward a state in which it understands itself, and we are part of its ongoing quest. This quest began when natural selection favored species that made correct predictions about their environment, moved on to organisms that became increasingly better at understanding nature, and now continues with our (more or less) organized scientific enterprise, nationally and internationally, individually and institutionally.

But what is this understanding we work toward? Understanding something means we are able to hold a workable model of it in our head, a simplified version of the real thing that we can question and that explains some aspect of what we observe. In physics, models are often heavily mathematical, and without lengthy training—for which not everybody has the time—it is impossible to fully grasp their properties. But once we have the mathematics, and at least *someone* understands it, it is often possible to communicate it verbally and visually. This book is my own little contribution to help you hold part of the universe in your head, using words and images rather than equations. By passing on knowledge, like my mother, I do my own part to aid the universe's understanding of itself.

So, yes, we are bags of atoms crawling around on a pale blue dot in the outer spiral arm of a remarkably unremarkable galaxy. And yet we are so much more than this.

Acknowledgments

First of all, I want to thank David Deutsch, Zeeya Merali, Tim Palmer, and Roger Penrose, whose interviews livened up this book. You were awesome. I also want to thank my agent, Max Brockman, and the wonderful people at Brockman Inc. who have supported me through the years, and my editor, Paul Slovak, and his team at Penguin Random House, who have done a great job getting this book into your hands.

Big thanks also to Timothy Gowers and Massimo Pigliucci for their help with parts of this book, and to John Horgan, Tim Palmer, Stefan Scherer, and Renate Weineck for reading early drafts of the manuscript.

For this book, I have drawn on more than ten years' experience answering queries from readers of my blog and viewers of my YouTube channel. They taught me to put aside technical terms and helped me to understand where non-experts have trouble following physicists' arguments. I am greatly indebted to their feedback. Above all, my audience served as a constant reminder that knowledge matters, regardless of whether it has technological applications. If you were one of them, my heartfelt thanks to you.

Existential Physics is dedicated to my husband, Stefan, who has suffered through more than his fair share of long-distance relationship. Over the twenty years we have known each other, he has patiently endured dozens of twists and turns in my whatabouts and whereabouts, and yet we somehow miraculously managed to get married, stay married, and raise two reasonably normal children. Through all this time, Stefan has been unwavering in his encouragement and support. For all this and so much more, thank you.

Sabine Hossenfelder,
Heidelberg, July 2021

Glossary

anthropic principle
The anthropic principle states that the universe must be the way it is so it allows for humans to exist. The weak anthropic principle merely acknowledges that this is a constraint that the laws of nature must fulfill, for otherwise they would be in conflict with evidence. The strong anthropic principle postulates in addition that the existence of humans *explains* why the laws of nature are the way they are.

classical
A classical theory is one that does not have quantum properties.

concordance model
Describes the universe on large scales. It includes all known types of matter in a **classical** approximation, and adds **dark matter** and **dark energy**. It uses the mathematical framework of **general relativity**. The concordance model is both **deterministic** and **local**. The concordance model is also known as ΛCMD (where Λ is the **cosmological constant** and CDM stands for "**cold dark matter**").

cosmological constant
A constant of nature, denoted Λ (capital lambda), that determines how fast the expansion of the universe accelerates. It's the simplest type of **dark energy** and makes up about 75 percent of the universe's matter-energy budget.

dark energy
Dark energy is a hypothetical type of energy that accelerates the expansion of the universe. Its simplest form is the **cosmological constant**.

GLOSSARY

dark matter
Dark matter is a hypothetical form of matter that makes up about 80 percent of the matter of the universe, or about 20 percent of the matter-energy budget. The observational evidence for dark matter is solid, but what it is made of (if anything) is unclear. Not to be confused with **dark energy**.

determinism, deterministic
A theory is deterministic if any given **initial condition** allows one to deduce the state of the system for all later times. **Classical** chaos is deterministic, and so is **general relativity**. The opposite of **non-deterministic**.

effective model
An effective model is an approximate description of a system at a desired level of resolution. All effective models are **emergent**. They are not merely emergent, however. They discard information deemed irrelevant for the purposes at hand.

emergent
An object, property, or law is emergent if it cannot be found or defined on the level of constituents and their behavior. If the emergent object, property, or law can be derived from the behavior and properties of the constituents, it is weakly emergent. If it cannot be so derived, it is strongly emergent. There are no known examples of strong emergence in nature.

evolution law
The evolution law is applied to the **initial state** of a system and allows us to calculate the state of the system at any later time. If the evolution law is **time-reversible**, we can also use it to calculate the state at any earlier time. All currently known evolution laws in the **foundations of physics** are differential equations.

foundations of physics
The research areas of physics concerned with **fundamental** laws. Those areas currently include high-energy particle physics, quantum gravity, quantum foundations, and parts of cosmology and astrophysics.

fundamental
A law, property, or object is fundamental if it cannot be derived from anything else. Fundamental is the opposite of **emergent**.

GLOSSARY

general relativity
Albert Einstein's theory of gravity, according to which gravity is the effect of a curved space-time. General relativity is **classical**, **local**, and **deterministic**. It is currently **fundamental**, but because of its incompatibility with **quantum field theory**, widely believed to be **emergent** from a more fundamental theory yet to be found.

inflation
A hypothetical phase of accelerated expansion in the early universe, conjectured to be created by a field called the *inflaton*. There is no convincing evidence for either inflation or the inflaton.

initial condition / initial state
Complete information about the state of a system at one particular moment in time, to which the **evolution law** is then applied. The state of the system in the initial condition is called the **initial state**.

local, locality
A theory is local if information transfer in this theory obeys the speed-of-light limit and if information, to go from one point to another, has to pass through all closed surfaces dividing these points. I want to warn the reader that physicists use several different definitions of *local*; this is only one of them, sometimes more specifically referred to as Einstein local. If you have heard that **quantum mechanics** is **nonlocal**, this was using a *different* notion of locality. In the definition used here, quantum mechanics *is* local, and so are the **standard model of particle physics** and the **concordance model**.

non-deterministic
A theory in which the later state of a system cannot be deduced from the **initial state** by the **evolution law**. It is the opposite of **deterministic**. A non-deterministic theory is also not **time-reversible**, but the opposite is not necessarily the case (a deterministic theory might not be time-reversible).

nonlocal
A theory in which spatially separated places can exchange information instantaneously. None of the currently known **fundamental** theories have this property. It is the opposite of **local**.

quantum field theory
A more complicated version of **quantum mechanics** in which particles interact by means of other particles. Like quantum mechanics, quantum field theory is **local** but **non-deterministic** and not **time-reversible**.

quantum mechanics
The theory by which we describe the behavior of particles (this includes light, which is made of particles called *photons*). Quantum mechanics is **local** but **non-deterministic** and not **time-reversible**.

reductionism
The practice of seeking better explanations by deriving an already known theory from a simpler theory. The theory that can be derived is then said to be *reducible* and the theory that it can be derived from is considered more **fundamental**. If the fundamental theory describes nature on shorter distances than the reducible theory, one often specifically speaks of *ontological reductionism*, whereas in the general case, one speaks of *theory reductionism*. Theory reductionism does not necessarily entail ontological reductionism, though historically they have gone hand in hand.

standard model of particle physics
The standard model describes the properties and behavior of all the experimentally confirmed particles and forces, except gravity, which is described by **general relativity**. It is a type of **quantum field theory** and therefore both **local** and **non-deterministic**. The standard model is currently **fundamental**.

time-reversible, time-reversibility
An **evolution law** is time-reversible if it maps one **initial state** to exactly one state at any other time. In this case, one can use the evolution law both forward and backward in time. The theory of **general relativity** is **time-reversible** in the absence of singularities. **Quantum field theories** are time-reversible except for the measurement process. A time-reversible theory is also **deterministic**, but a deterministic theory is not necessarily time-reversible.

Notes

Preface

ix **"It is far better":** Carl Sagan, *The Demon-Haunted World: Science as a Candle in the Dark* (New York: Ballantine Books, 1997), 12.

xiv **"insights into turgid prose":** Nicholas Kristof, "Professors, We Need You!," *New York Times*, Feb. 14, 2014.

Chapter 1

3 **if you run in a circle:** Acceleration is a change of velocity. Both acceleration and velocity are vectors, meaning they have a direction. A change of direction of velocity is therefore also an acceleration, even if the magnitude of the velocity (the speed) remains constant.

4 **too far off topic:** A lovely introduction to special relativity is Chad Orzel's *How to Teach Relativity to Your Dog* (New York: Basic Books, 2012). If you want to see more of the math, a good starting point is Leonard Susskind and Art Friedman, *Special Relativity and Classical Field Theory: The Theoretical Minimum* (New York: Basic Books, 2017).

11 **agree that *anything* exists now:** Let me be clear that I am not trying to tell you what it means for something to exist in the first place. This is arguably a tricky question. The argument is rather a statement about what, according to special relativity, exists the same way. You can, for example, circumvent the conclusion by arguing that nothing exists that isn't in the same place as you, so that light doesn't have to travel for you to see it. Leaving aside that this, strictly speaking, means that the only things that exist are in your brain, it just isn't how most of us use the word *exist*.

11 **To sum it up:** John Lloyd, in "The Infinite Money Cage: Parallel Universes," BBC Radio 4, July 16, 2012.

12 **discovery of quantum mechanics:** Or at least believed to be deterministic. There are some subtle cases in which Newtonian mechanics becomes indeterministic, but Laplace didn't know that.

12 **In 1814, the French scientist:** Pierre-Simon Laplace, *Essai philosophique sur les probabilités* (Paris: Courcier, 1814; repr. Paris: Forgotten Books, 2018). Trans. Frederick Wilson Truscott and Frederick Lincoln Emory (New York: Wiley, 1902).

16 **"nobody understands quantum mechanics":** Richard Feynman, "Messenger Lectures at Cornell: The Character of Physical Law, Part 6: Probability and Uncertainty" (1964). You can find a recording on YouTube here: youtube.com/watch?v=Ja0HSFj8Imc. The quote is at about eight minutes into the lecture.

16–17 **"even physicists don't understand":** Sean Carroll, "Even Physicists Don't Understand Quantum Mechanics," *New York Times*, Sept. 7, 2019.

17 **If you want to know more:** Adam Becker, *What Is Real? The Unfinished Quest for the Meaning of Quantum Physics* (New York: Basic Books, 2018); Philip Ball, *Beyond Weird: Why Everything You Thought You Knew about Quantum Physics Is Different* (Chicago: University of Chicago Press, 2018); and Jim Baggott, *Quantum Reality: The Quest for the Real Meaning of Quantum Mechanics—a Game of Theories* (New York: Oxford University Press, 2020).

NOTES

20 **"unreasonably effective"**: Eugene Wigner, "The Unreasonable Effectiveness of Mathematics in the Natural Sciences," *Communications on Pure and Applied Mathematics* 13 (1960): 1–14.

Chapter 2

25 **Cosmology is one of the cases:** For an example of the type of analysis I have in mind, see Debika Chowdhury, Jérôme Martin, Christophe Ringeval, and Vincent Vennin, "Assessing the Scientific Status of Inflation after Planck," *Physical Review D* 100, no. 8 (Oct. 24, 2019): 083537, arXiv:1902.03951 [astro-ph.CO].

25 **This can be done:** Not only one, but a bewildering variety. The research area is called *morphometrics*, and if you want to know more, Wikipedia is a good entry point.

32 **this claim has been contested:** Anna Ijjas and Paul J. Steinhardt, "Implications of Planck 2015 for Inflationary, Ekpyrotic and Anamorphic Bouncing Cosmologies," *Classical and Quantum Gravity* 33 (2016): 044001, arXiv:1512.09010 [astro-ph.CO].

32 **creation ex nihilo:** Lawrence Krauss, *A Universe from Nothing: Why There Is Something Rather than Nothing* (New York: Free Press, 2012).

33 **It has since disappeared:** Niayesh Afshordi, Daniel J. H. Chung, and Ghazal Geshnizjani, "Cuscuton: A Causal Field Theory with an Infinite Speed of Sound," *Physical Review D* 75 (2007): 083513, arXiv:hep-th/0609150.

33 **can't be distinguished from inflation:** Ghazal Geshnizjani, William H. Kinney, and Azadeh Moradinezhad Dizgah, "General Conditions for Scale-Invariant Perturbations in an Expanding Universe," *Journal of Cosmology and Astroparticle Physics* 11 (2011): 049.

35 **This network then changes:** Tomasz Konopka, Fotini Markopoulou, and Simone Severini, "Quantum Graphity: A Model of Emergent Locality," *Physical Review D* 77 (2008): 104029, arXiv:0801.0861 [hep-th].

38 **the laws of nature:** David Hume, *A Treatise of Human Nature*, ed. Lewis Amherst Selby-Bigge (Oxford, UK: Clarendon Press, 1896).

38–39 **laws of living on a farm:** Bertrand Russell, *The Problems of Philosophy* (New York: Barnes & Noble, 1912).

39 **97 percent of all Wikipedia:** "Wikipedia:Getting to Philosophy," Wikimedia Foundation, last modified December 29, 2021, 01:14, en.wikipedia.org/wiki/Wikipedia:Getting_to_Philosophy.

Other Voices #1

45 **doesn't increase the banana harvest:** Liam Fox, "Bananas-for-Sex Cult Leader on the Run," abc.net.au, Sept. 15, 2009.

45 **rather than proclaiming:** Meredith Bennett-Smith, "Lawrence Krauss, Physicist, Claims Teaching Creationism Is Child Abuse and 'Like the Taliban,'" *HuffPost*, Feb. 14, 2013.

46 **Hawking in his book:** Stephen Hawking, *A Brief History of Time* (New York: Bantam Books, 1988).

Chapter 3

56 **the past-hypothesis:** The fact that the universe requires an initial condition of low entropy to reproduce our observations was discussed by physicists already in the early days of thermodynamics, but the term *past-hypothesis* was coined much later by David Albert in his book *Time and Chance* (Cambridge, MA: Harvard University Press, 2000).

56 **Penrose's conformal cyclic cosmology:** Roger Penrose, *Cycles of Time: An Extraordinary New View of the Universe* (London: Bodley Head, 2010).

56 **Sean Carroll thinks:** Sean Carroll, *From Eternity to Here: The Quest for the Ultimate Theory of Time* (New York: Penguin, 2010).

56 **And Julian Barbour:** Julian Barbour, *The Janus Point: A New Theory of Time* (London: Bodley Head, 2020).

61 **The super-nerds:** You can find out more about the Euler-Mascheroni constant in Julian Havil, *Gamma: Exploring Euler's Constant* (Princeton University Press, 2003).

62 **But change the notion of:** David Bohm, *Wholeness and the Implicate Order* (Abingdon, UK: Routledge, 1980). I don't think this is how Bohm himself understood the terms *implicate order* and *explicate order*. However, I believe the way I used them to distinguish easily discernible from hidden differences is close to what he had in mind.

NOTES

64 **"The consciousness of AC"**: Isaac Asimov, "The Last Question," *The Best of Isaac Asimov* (Garden City, NY: Doubleday, 1974).

64 **"The problem of the Now"**: Rudolf Carnap, "Intellectual Autobiography," in Paul Arthur Schilpp, ed., *The Philosophy of Rudolf Carnap* (Chicago: Open Court, 1963).

65 **replacing the current theory**: Fay Dowker, "Causal Sets and the Deep Structure of Spacetime," in Abhay Ashtekar, ed., *100 Years of Relativity—Space-time Structure: Einstein and Beyond* (Singapore: World Scientific, 2005), arXiv:gr-qc/0508109.

65 **revision of quantum mechanics**: N. David Mermin, "Physics: QBism Puts the Scientist Back into Science," *Nature* 507 (2014): 421–23.

65 **mathematics itself is the problem**: Lee Smolin, "The Unique Universe," *Physics World*, June 2, 2009, physicsworld.com/a/the-unique-universe.

69 **"Schrödinger's Cat"**: Lyrics from a song I wrote some years ago: youtube.com/watch?v=I_0IaAhvHKE.

71 **temporarily defies the second law**: G. M. Wang et al., "Experimental Demonstration of Violations of the Second Law of Thermodynamics for Small Systems and Short Time Scales," *Physical Review Letters* 89, no. 5 (Aug. 2002): 050601.

72 **The physicist Seth Lloyd**: Quoted in Lisa Grossman, "Quantum Twist Could Kill Off the Multiverse," *New Scientist*, May 14, 2014.

72 **I side with Sean**: Sean M. Carroll, "Why Boltzmann Brains Are Bad," in Shamik Dasgupta, Ravit Dotan, and Brad Weslake, eds., *Current Controversies in Philosophy of Science* (London: Routledge, 2020), arXiv:1702.00850 [hep-th].

75 **Is it really a rabbit**: Thomas Kuhn used the same example in his book *The Structure of Scientific Revolutions* (Chicago: University of Chicago Press, 1962) to illustrate a paradigm change. This is not what I am referring to here.

Chapter 4

80 **The heaviest of the elements**: For a long time, astrophysicists thought the heaviest elements were created in supernova explosions, but according to the most recent data, the better hypothesis is that the heavy elements are created in neutron star mergers. See, e.g., Darach Watson et al., "Identification of Strontium in the Merger of Two Neutron Stars," *Nature* 574 (Oct. 2019): 497–500.

81 **a hundred trillion years**: Fred Adams and Greg Laughlin, *The Five Ages of the Universe* (New York: Free Press, 1999).

81 **In a 2019 survey**: David Wisniewski, Robert Deutschländer, and John-Dylan Haynes, "Free Will Beliefs Are Better Predicted by Dualism Than Determinism Beliefs across Different Cultures," *PLOS ONE* 14, no. 9 (Sept. 11, 2019): e0221617.

84 **Didn't Philip Anderson**: Philip W. Anderson, "More Is Different," *Science* 177, no. 4047 (Aug. 4, 1972): 393–96.

85 **many other effective models**: For a technical introduction, see, e.g., C. P. Burgess, "Introduction to Effective Field Theory," *Annual Review of Nuclear and Particle Science* 57 (2007): 329–62, arXiv:hep-th/0701053.

88 **an abundance of similar examples**: There are a few other supposed counterexamples that people have put forward to me—for example, global conditions like boundary values, or topological constraints. But those can all be defined in microscopic terms. Again, if you want to show that reductionism fails, you'll have to find an example that cannot be derived from the microphysics. I have discussed this in more detail in Sabine Hossenfelder, "The Case for Strong Emergence," in Anthony Aguirre, Brendan Foster, and Zeeya Merali, eds., *What Is Fundamental?* (New York: Springer, 2019), 85–94.

90 **replaced every two weeks**: Kirsty L. Spalding et al., "Retrospective Birth Dating of Cells in Humans," *Cell* 122, no. 1 (Aug. 2005): 133–43.

91 **thought experiment in 1980**: Zenon W. Pylyshyn, "Computation and Cognition: Issues in the Foundations of Cognitive Science," *Behavioral and Brain Sciences* 3, no. 1 (Mar. 1980): 111–69.

93 **physicist and Nobel Prize winner**: Gerard 't Hooft, *The Cellular Automaton Interpretation of Quantum Mechanics* (New York: Springer, 2016).

NOTES

Other Voices #2

96 **I am here because:** David Deutsch, *The Fabric of Reality: The Science of Parallel Universes—and Its Implications* (New York: Viking, 1997) and *The Beginning of Infinity: Explanations That Transform the World* (New York: Penguin, 2011).

98 **So if this is a fundamental:** For a more detailed exposition of Turing computability, see David Deutsch, "Quantum theory, the Church-Turing principle and the universal quantum computer," The Royal Society. A40097–117 (1985).

99 **The causal exclusion principle then:** Jaegwon Kim, "Making Sense of Emergence," *Philosophical Studies* 95, no. 1–2 (Aug. 1999): 3–36; and "Emergence: Core Ideas and Issues," *Synthese* 151, no. 3 (Aug. 2006): 547–59.

Chapter 5

105 **"an irreconcilable mismatch":** Anil Ananthaswamy, "Spin-Swapping Particles Could Be 'Quantum Cheshire Cats,'" *Scientific American*, May 6, 2019; and George Musser, "Quantum Paradox Points to Shaky Foundations of Reality," *Science*, Aug. 17, 2020.

107 **I agree with Philip Ball:** Philip Ball, *Beyond Weird: Why Everything You Thought You Knew about Quantum Physics Is Different* (Chicago: University of Chicago Press, 2018).

109 **"spooky action at a distance":** Albert Einstein, letter to Max Born on March 3, 1947, in *Albert Einstein Max Born Briefwechsel 1916–1955* (Munich: Nymphenburger Verlangshandlung, 1991).

109 **one can mathematically prove:** It's called the *non-signaling theorem* or *no-communication theorem* and can be found in most textbooks and on Wikipedia. It goes back to Giancarlo Ghirardi, Alberto Rimini, and Tullio Weber, "A General Argument against Superluminal Transmission through the Quantum Mechanical Measurement Process," *Lettere al Nuovo Cimento* 27, no. 10 (1980): 293–98.

109 **This is a possibility:** Sabine Hossenfelder and Tim Palmer, "How to Make Sense of Quantum Physics," *Nautilus*, Mar. 12, 2020; and "Rethinking Superdeterminism," *Frontiers in Physics* 8 (May 6, 2020): 139, arXiv:1912.06462.

113 **the Universe Splitter:** apps.apple.com/us/app/universe-splitter/id329233299.

118 **On September 22, 2015:** Dave Levitan, "Carson rewrites laws of thermodynamics," *Philadelphia Inquirer*, Sept. 25, 2015, inquirer.com/philly/news/politics/factcheck/SciCheck_Carson _rewrites_laws_of_thermodynamics.html.

118 **In an earlier speech:** "Ben Carson in 2012 speech: The Big Bang Is a Fairytale," youtube .com/watch?v=DJo7R0OfC5M.

118 **misunderstood much:** Lawrence Krauss commented on Carson's speech and explained the mistakes in Lawrence Krauss, "Ben Carson's Scientific Ignorance," *New Yorker*, Sept. 28, 2015.

119 **The simulation hypothesis:** Nick Bostrom, "Are You Living in a Computer Simulation?" *Philosophical Quarterly* 53, no. 211 (Apr. 2003): 243–55.

119 **Elon Musk is among those:** Elon Musk, in "Joe Rogan & Elon Musk—Are We in a Simulated Reality?," Sept. 7, 2018, youtube.com/watch?v=0cM690CKArQ.

119 **even Neil deGrasse Tyson:** Corey S. Powell, "Elon Musk Says We May Live in a Simulation. Here's How We Might Tell If He's Right," NBC News, Oct. 2, 2018.

121 **conventional computer in finite time:** Zohar Ringel and Dmitri L. Kovrizhin, "Quantized Gravitational Responses, the Sign Problem, and Quantum Complexity," *Science Advances* 3, no. 9 (Sept. 2017): e1701758.

121 **Indeed, physicists have looked:** Silas R. Beane, Zohreh Davoudi, and Martin J. Savage, "Constraints on the Universe as a Numerical Simulation," *European Physical Journal A* 50, no. 9 (Oct. 2012): 148.

Chapter 6

125 **"garden of forking paths":** Jorge Luis Borges, *The Garden of Forking Paths* (New York: Penguin, 2018); original: "El jardín de senderos que se bifurcan," (Buenos Aires: Sur, 1941).

127 **As Ludwig Wittgenstein put it:** Ludwig Wittgenstein, *Logisch-philosophische Abhandlung [Tractatus Logico-Philosophicus]* (London: Kegan Paul, 1922).

128 **59 percent identified as compatibilists:** philpapers.org/surveys.

NOTES

128 a little more about compatibilism: Immanuel Kant, *The Critique of Practical Reason*, ed. Mary J. Gregor (New York: Cambridge University Press, 1997); William James, "The Dilemma of Determinism," *Unitarian Review*, Sept. 1884, in *The Will to Believe* (New York: Dover, 1956); and Wallace I. Matson, "On the Irrelevance of Free-Will to Moral Responsibility, and the Vacuity of the Latter," *Mind* 65, no. 260 (Oct. 1956): 489–97.

128 The philosopher John Martin Fischer: John Martin Fischer et al., *Four Views on Free Will* (Hoboken, NJ: Wiley-Blackwell, 2007).

129 The philosopher Jenann Ismael: Jenann Ismael, *How Physics Makes Us Free* (New York: Oxford University Press, 2016).

130 A pepped-up version: Philip Ball, "Why Free Will Is beyond Physics," *Physics World*, Jan. 2021.

130 When Sean Carroll: Sean Carroll, "Free Will Is as Real as a Baseball," *Discover*, July 13, 2011.

131 A 2019 survey: Ivar R. Hannikainen et al., "For Whom Does Determinism Undermine Moral Responsibility? Surveying the Conditions for Free Will across Cultures," *Frontiers in Psychology* 10 (Nov. 5, 2019): 2428.

131 often rely on certain approximations: John F. Donoghue, "When Effective Field Theories Fail," *Proceedings of Science, International Workshop on Effective Field Theories* 09, 001 (Feb. 2–6, 2009), arXiv:0909.0021 [hep-ph].

132 What gets us a little closer: Mile Gu et al., "More Really Is Different," *Physica D: Nonlinear Phenomena* 238, no. 9–10 (May 2009): 835–39, arXiv:0809.0151 [cond-mat.other]; and Toby S. Cubitt, David Perez-Garcia, and Michael M. Wolf, "Undecidability of the Spectral Gap," *Nature* 528, no. 7581 (Dec. 2015): 207–11.

132 It's a long shot: I have laid this out in more detail in Sabine Hossenfelder, "The Case for Strong Emergence," in Anthony Aguirre, Brendan Foster, and Zeeya Merali, eds., *What Is Fundamental?* (New York: Springer, 2019), 85–94.

136 Montgomery was diagnosed: Rachel Louise Snyder, "Punch after Punch, Rape after Rape, a Murderer Was Made," *New York Times*, Dec. 18, 2020.

139 This view was expressed: Azim F. Shariff and Kathleen D. Vohs, "What Happens to a Society That Does Not Believe in Free Will?," *Scientific American*, June 1, 2014.

139 For example, a 2017 study: Emilie A. Caspar et al., "The Influence of (Dis)belief in Free Will on Immoral Behavior," *Frontiers in Psychology* 8, article 20 (Jan. 17, 2017).

139 For the no-free-will priming: Francis Crick, *The Astonishing Hypothesis: The Scientific Search for the Soul* (New York: Scribner, 1995).

Other Voices #3

147 quantum states of the microtubules: Stuart Hameroff, "How Quantum Brain Biology Can Rescue Conscious Free Will," *Frontiers in Integrative Neuroscience* 6 (Oct. 2012): 93.

147 *orchestrated objective reduction*: Stuart Hameroff and Roger Penrose, "Consciousness in the Universe: A Review of the 'Orch OR' Theory,"*Physics of Life Reviews* 11, no. 1 (Mar. 2014): 39–78.

147 The major reason: Max Tegmark, "Importance of Quantum Decoherence on Brain Processes," *Physical Review E* 61, no. 4 (May 2000): 4194–206, arXiv:quant-ph/9907009.

147 it'd take a significant modification: Stuart Hameroff and Roger Penrose, "Reply to Seven Commentaries on 'Consciousness in the Universe: Review of the "Orch OR" Theory,'" *Physics of Life Reviews* 11, no. 1 (Dec. 2013): 94–100.

Chapter 7

152 The currently known laws: John Baez, "How Many Fundamental Constants Are There?," University of California–Riverside, College of Natural and Agricultural Sciences, Department of Mathematics, Apr. 22, 2011, math.ucr.edu/home/baez/constants.html.

153 a universe for any possible combination: Particle physicists use the same type of argument when they ask for the next larger particle collider. In that case, they claim it requires an explanation of why the mass of the Higgs boson happens to be what it is. This is called an *argument from naturalness*. I explain this in detail in my book *Lost in Math: How Beauty Leads Physics Astray* (New York: Basic Books, 2018).

154 to find out how likely it is: Luke A. Barnes et al., "Galaxy Formation Efficiency and the Multiverse Explanation of the Cosmological Constant with EAGLE Simulations," *Monthly Notices of the Royal Astronomical Society* 477, no. 3 (Jan. 2018).

155 **"[probability distributions]"**: The actual term in the paper is *measure*. A measure generally gives weight to an abstract space—for example, the space of all possible combinations of constants. For the purposes of the present discussion, it means the same as probability distribution.

157 **My discussion partner**: Geraint F. Lewis and Luke A. Barnes, *A Fortunate Universe: Life in a Finely Tuned Cosmos* (Cambridge, UK: Cambridge University Press, 2016).

158 **could have been anything**: The word *anything*, strictly speaking, is not correct, because a probability distribution over an infinite range of values cannot be normalized to 1. Strictly speaking, it should be "could take on values distributed over many orders of magnitude." It doesn't really matter, though. The point is that the prior, whatever it is, can't be justified.

158 **attempt to prove that God exists**: Dan Kopf, "The Most Important Formula in Data Science Was First Used to Prove the Existence of God," *Quartz*, June 30, 2018.

161 **nice thing about path integrals**: It would lead us somewhat astray to go into this in detail, but all these possible interactions can diagrammatically be represented by graphs, commonly called *Feynman diagrams*. This is very nicely explained in Gavin Hesketh, *The Particle Zoo: The Search for the Fundamental Nature of Reality* (London: Quercus, 2016).

162 **Or no constants at all**: One can quibble with the number 26, because it doesn't include some constants that could be there but that we simply set to zero because we've never observed anything contradicting this value. The mass of the elementary particle called the *gluon*, for example, is usually just set to zero because we have no experimental evidence that suggests otherwise. Yet one could add these masses as free parameters too. Strictly speaking, then, there are infinitely many possible constants that we set to zero. Another way to say this is that it's difficult to tell apart constants from the equations in which they appear. Alas, this is all rather irrelevant for the question whether and how our current theories can be further simplified.

164 **I leave you references**: Some references for constants of nature that are nothing like our own yet give rise to complex chemistry: Roni Harnik, Graham D. Kribs, and Gilad Perez, "A Universe without Weak Interactions," *Physical Review D* 74 (Aug. 17, 2006): 035006, arXiv:hep-ph/0604027; Fred C. Adams and Evan Grohs, "Stellar Helium Burning in Other Universes: A Solution to the Triple Alpha Fine-Tuning Problem," *Astroparticle Physics* 87 (Aug. 2016), arXiv:1608.04690 (astro-ph.CO); Abraham Loeb, "The Habitable Epoch of the Early Universe," *International Journal of Astrobiology* 13, no. 4 (Dec. 2013): 337–39, arXiv:1312.0613 (astro-ph.CO); and Don N. Page, "Preliminary Inconclusive Hint of Evidence against Optimal Fine Tuning of the Cosmological Constant for Maximizing the Fraction of Baryons Becoming Life" (Jan. 2011), arXiv:1101.2444 [hep-th].

164 *cosmological natural selection*: Lee Smolin, *The Life of the Cosmos* (New York: Oxford University Press, 1998).

Chapter 8

169 **200 billion galaxies**: Tod R. Lauer et al., "New Horizons Observations of the Cosmic Optical Background," *Astrophysical Journal* 906, no. 2 (Jan. 2021): 77, arXiv:2011.03052 [astro-ph.GA].

170 **the human brain and the universe**: Franco Vazza and Alberto Feletti, "The Quantitative Comparison between the Neuronal Network and the Cosmic Web," *Frontiers in Physics* 8 (2020): 525731.

176 **Einstein indeed used this phrase**: Albert Einstein, letter to Max Born on March 3, 1947, in *Albert Einstein Max Born Briefwechsel 1916–1955* (Munich: Nymphenburger Verlangshandlung, 1991).

176 **As Penrose pointed out**: This topic has been explored in depth in George Musser, *Spooky Action at a Distance* (New York: Scientific American/Farrar, Straus and Giroux, 2015).

177 **There may be only normal matter**: Friedrich W. Hehl and Bahram Mashhoon, "Nonlocal Gravity Simulates Dark Matter," *Physics Letters B* 673, no. 4–5 (Jan. 2009): 279–82, arXiv:0812.1059 [gr-qc].

178 **defects might have been left**: Fotini Markopoulou and Lee Smolin, "Disordered Locality in Loop Quantum Gravity States," *Classical and Quantum Gravity* 24, no. 15 (Mar. 2007): 3813–24, arXiv:gr-qc/0702044 [gr-qc].

NOTES

180 **The mathematician and entrepreneur:** John Horgan, "Polymath Stephen Wolfram Defends His Computational Theory of Everything," *Scientific American, Cross-Check* blog, Mar. 5, 2017.

182 **The physicist John Wheeler:** John Archibald Wheeler, *Relativity, Groups and Topology: Lectures Delivered at Les Houches During the 1963 Session of the Summer School of Theoretical Physics,* eds. Bryce DeWitt and Cécile DeWitt-Morette (New York: Gordon and Breach, 1964), 408–31.

183 **This idea has been promoted:** Deepak Chopra, "The Mystery That Makes Life Possible," Deepak Chopra.com, Oct. 24, 2020; Philip Goff, "Panpsychism Is Crazy, but It's Also Most Probably True," *Aeon,* Mar. 1, 2017; and Christof Koch, "Is Consciousness Universal?" *Scientific American Mind,* Jan. 1, 2014.

185 **It's incompatible with data:** The free will theorem (John Conway and Simon Kochen, "The Free Will Theorem," *Foundations of Physics* 36, no. 10 [Jan. 2006]: 1441–73, arXiv:quant-ph /0604079) plays no role whatsoever in this argument. In fact, the free will theorem has nothing to do with free will. It is merely about an assumption in another theorem that is sometimes, misleadingly, referred to as the *free will assumption.* Even if that weren't so, all that the theorem says is that (given certain assumptions) if humans have free will, then so do elementary particles. If the theorem was really about free will, the obvious conclusion from this would be that humans have no free will.

187 **It was put forward:** Giulio Tononi, "An Information Integration Theory of Consciousness," BioMed Central, *BMC Neuroscience* 5, no. 1 (Nov. 2004): 42.

188 **Estimates show that:** Carl Zimmer, "Sizing Up Consciousness by Its Bits," *New York Times,* Sept. 20, 2010.

188 **The magazine *New Scientist*:** Quoted in Michael Brooks, "Here. There. Everywhere?" *New Scientist* 246, no. 3280 (May 2, 2020): 40–44. By the time of writing (May 2021), the work that Bor was referring to has unfortunately still not appeared.

188 **Yet another problem for IIT:** Scott Aaronson, "Why I Am Not an Integrated Information Theorist (or, the Unconscious Expander)," *Shtetl-Optimized* blog, May 21, 2014, scottaaronson .com/blog/?m=201405.

188 **There are some other measures:** Jose L. Perez Velazquez, Diego M. Mateos, and Ramon Guevara Erra, "On a Simple General Principle of Brain Organization," *Frontiers in Neuroscience* 13 (Oct. 15, 2019): 1106; and Sophia Magnúsdóttir, "I Think, Therefore I Think You Think I Am," in Anthony Aguirre, Brendan Foster, and Zeeya Merali, eds., *Wandering Towards a Goal: The Frontiers Collection* (New York: Springer, 2018).

189 **Mary's room is:** Frank Jackson, "Epiphenomenal Qualia," *Philosophical Quarterly* 32, no. 127 (Apr. 1982): 127–36.

189 **Jackson himself later abandoned:** Frank Jackson, "Postscript on Qualia," *Mind, Method, and Conditionals: Selected Essays* (London: Routledge, 1998), 76–79.

Other Voices #4

191 **In 2017, Zeeya:** Zeeya Merali, *A Big Bang in a Little Room: The Quest to Create New Universes* (New York: Basic Books, 2017). Zeeya spent so much time talking to others about the topic, she later wrote a review article about it: Stefano Ansoldi, Zeeya Merali, and Eduardo I. Guendelman, "From Black Holes to Baby Universes: Exploring the Possibility of Creating a Cosmos in the Laboratory," *Bulgarian Journal of Physics* 45, no. 2 (Jan. 2018): 203–20, arXiv:1801.04539 [gr-qc].

Chapter 9

199 **93 percent predictable:** Chaoming Song et al., "Limits of Predictability in Human Mobility," *Science* 327, no. 5968 (Feb. 2010): 1018–21.

200 **Scriven claimed:** Michael Scriven, "An Essential Unpredictability in Human Behaviour," in Benjamin B. Wolman and Ernest Nagel, eds., *Scientific Psychology: Principles and Approaches* (New York: Basic Books, 1965), 411–25.

201–202 **the *domino problem*:** Hao Wang, "Proving Theorems by Pattern Recognition—II," *Bell System Technical Journal* 40, no. 1 (Jan. 1961): 1–41.

NOTES

202 **This was proved in 1966:** Robert Berger, "The Undecidability of the Domino Problem," *Memoirs of the American Mathematical Society* 66 (Providence, RI: American Mathematical Society, 1966).

202 **Turing's halting problem:** Alan Turing, "On Computable Numbers, with an Application to the Entscheidungsproblem," *Proceedings of the London Mathematical Society*, series 2, no. 42 (1937): 230–65.

204 **That way, Gödel's theorem:** Lawrence C. Paulson, "A Mechanised Proof of Gödel's Incompleteness Theorems Using Nominal Isabelle," *Journal of Automated Reasoning* 55, no. 1 (June 2015): 1–37.

204 **find quite charming the argument:** In the notation of the entry on "Gödel's Incompleteness Theorem," *Stanford Encyclopedia of Philosophy*, plato.stanford.edu/entries/goedel -incompleteness.

204 **Indeed, chaos could:** Tim N. Palmer, Andreas Döring, and Gregory Seregin,"The Real Butterfly Effect," *Nonlinearity* 27, no. 9 (Aug. 2014): R123.

205 **differential equations with this behavior:** John M. Ball, "Finite Time Blow-up in Nonlinear Problems," in *Nonlinear Evolution Equations: Proceedings of a Symposium Conducted by the Mathematics Research Center, the University of Wisconsin–Madison, October 17–19, 1977,* Michael G. Crandall, ed. (Cambridge, MA: Academic Press, 1978), 189–205.

205 **Quantum theories are not chaotic:** Quantum theory is linear in the wave function. Chaos requires a nonlinear theory. The Lagrangian is usually nonlinear in field operators, but those have to be evaluated for a particular wave function. The research area of quantum chaos uses a definition of *chaos* that differs from the one used in other fields.

205 **number four on the list:** Clay Mathematics Institute, "Millennium Problems," claymath.org /millennium-problems.

208 **Elon Musk thinks:** Quoted in Samuel Gibbs, "Elon Musk: Artificial Intelligence Is Our Biggest Existential Threat," *Guardian*, Oct. 27, 2014.

208 **Stephen Hawking said:** Quoted in Arjun Kharpal, "Stephen Hawking Says A.I. Could Be 'Worst Event in the History of Our Civilization,'" CNBC, Nov. 6, 2017.

208 **Apple cofounder Steve Wozniak:** Quoted in Peter Holley, "Apple Co-founder on Artificial Intelligence: 'The Future Is Scary and Very Bad for People,'" *Washington Post*, Mar. 24, 2015.

208 **And Bill Gates:** Quoted in Peter Holley, "Bill Gates on Dangers of Artificial Intelligence: 'I Don't Understand Why Some People Are Not Concerned,'" *Washington Post*, Jan. 29, 2015.

208 **In 2015, the Future of Life Institute:** Stuart Russell, Daniel Dewey, and Max Tegmark, "Research Priorities for Robust and Beneficial Artificial Intelligence," *AI Magazine* (Winter 2015): 105–14, Association for the Advancement of Artificial Intelligence at Future of Life Institute, futureoflife.org/data/documents/research_priorities.pdf?x40372.

208 **a body of some kind:** Carlos E. Perez, "Embodied Learning Is Essential to Artificial Intelligence," Intuition Machine, Medium.com, Dec. 12, 2017.

213 **not just the fantasy of a theorist:** Julian Kates-Harbeck, Alexei Svyatkovskiy, and William Tang, "Predicting Disruptive Instabilities in Controlled Fusion Plasmas Through Deep Learning," *Nature* 568, no. 7753 (Apr. 2019): 526–31.

214 **research line of *computable economics*:** K. Vela Velupillai, "Towards an Algorithmic Revolution in Economic Theory," *Journal of Economic Surveys* 25, no. 3 (July 2011): 401–30.

215 **Even more interesting:** Tim N. Palmer, "Human Creativity and Consciousness: Unintended Consequences of the Brain's Extraordinary Energy Efficiency?" *Entropy* 22, no. 3 (Feb. 2020): 281, arXiv:2002.03738 [q-bio.NC].

Epilogue

219 **Stephen Jay Gould got it right:** Stephen J. Gould, "Nonoverlapping Magisteria," *Natural History* 106 (Mar. 1997): 16–22, 60–62.

Index

INDEX

INDEX